Contents

Outline of the scheme

The problems are subdivided into 61 topic areas, contained in two Student's Books. The 61 chapters of the Teacher's books support the problem units and show how the units relate to National Curriculum Attainment Targets. Each chapter of the Teacher's book contains

- A short introduction to the particular topic and how it can be approached from a problem-solving angle.
- A grid relating the problems to the National Curriculum.
- Comments and suggestions. These offer comments on the mathematics involved in each problem and suggestions about how they may be tackled in class. Other features include (where appropriate): historical information, hints on how to tackle the problems, and suggestions for further problems and investigations.

In addition, the author's introduction to Problem Solving in the Classroom appears at the beginning of Teacher's Book 1.

Two Solutions Packs offer:

- Solutions to the problems unit-by-unit. These complement the comments and suggestions of the Teacher's Book, and where the problem has a definite 'answer' this is given. The emphasis, however, is on different ways of looking at the problems and on the richness hidden within a problem rather than a single approach. Additional problems posed by the notes of the Teacher's Book are also commented on. Teachers and pupils may of course discover yet more problems to investigate as a result of these solutions.

For ordering purposes, ISBN's for all of these publications will be found on the back cover.

PROBLEM SOLVING FOR NATIONAL CURRICULUM MATHEMATICS

Teacher's Book 1

DAVID WELLS

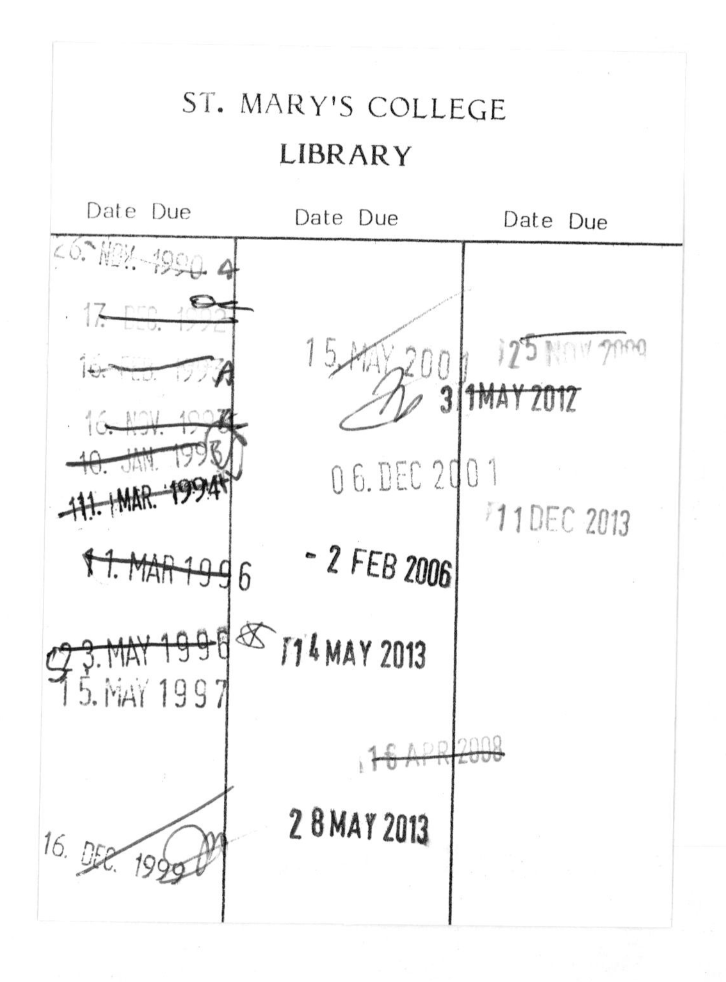

This material first published in *Mathematics through problem solving* 1988
Separate *Teacher's Book 1* first published 1990

Basil Blackwell Ltd
108 Cowley Road
Oxford OX4 1JF
England

British Library Cataloguing in Publication Data

Wells, D. G. (David Graham) *1940–*
Problem solving for national curriculum mathematics.
Teacher's book
1. Mathematics. Problem solving
I. Title
510

ISBN 0–631–90534–0

Design by David Chaundy
Illustrations by Anne Langford
Typeset by Unicus Graphics Ltd, Horsham
Printed in Great Britain by Dotesios (Printers) Ltd

Teaching the mathematics syllabus through problem solving

Introduction

The Cockcroft Report and the GSCE Criteria are only two of many recent documents which emphasise the importance of problem solving in the mathematical classroom. How is it to be introduced?

Problem solving can appear to pupils as an extra, as an addition to the normal work on the syllabus. As such, it introduces pupils to the activity of being creative mathematically, gives them insight into how professionals work and think, gives them opportunities to produce their own ideas and work independently, increases motivation in almost all pupils, and produces surprisingly successful results from pupils who perform rather poorly when taught directly by the teacher.

This is probably the best way to introduce problem solving. It requires some time, perhaps one or two periods each week, to be set aside for problem solving, but otherwise the normal classroom routine of the pupils and the teacher is not disturbed.

However, problem solving can, and I would argue should, have a much larger role than this. There is no reason why mathematics itself should not be taught through problem solving, broadly understood to include all the pupil's own mathematical activity, so that problem solving becomes not an extra but the bedrock on which all the teaching and learning in the classroom is founded.

Naturally in order to achieve this aim, the curriculum must be looked at in a new and different light. A topic on the written syllabus or a chapter in a textbook must be seen not as the subject for a single clear explanation, but as a starting point for an illuminating variety of possible problems and explorations.

The teacher's expectations of what a topic 'means' also need to change. The methods presented in the textbooks tend to be extremely powerful and general. Approached via problem solving, there are many other methods which will be illuminating for the pupils and which can be used to make the topic richer and more meaningful, on the way to the standard solutions which will be appropriate for pupils who progress that far.

The level of ability at which a topic can be tackled will also change. When pupils are allowed to tackle problems by the means which are natural to their age and ability they will often prove more able and successful than usual.

This is true of younger pupils, who can often tackle topics usually taught to older pupils, and weaker pupils, including pupils all the way down to the bottom stream and in the remedial department, who prove capable of much greater progress than when taught directly by their teacher.

The results of such an active approach can be greater intuitive understanding by the pupils and a much more firmly based understanding, as well as greater enjoyment and motivation.

Presenting the problems

A choice of problems and pupil responsibility

Some (rare) problems do allow all pupils in a mixed ability class to work at their own level, and to be stretched without being broken. (Remember in this context that in a five-stream school, only the middle of the five streams can conceivably claim *not* to be mixed ability.)

An example is: what closed shapes can you make by joining squares and equilateral triangles edge to edge? Most problems, however, do not pose equal difficulties to pupils at different levels, and therefore posing only one problem to a whole class will not suit them all.

The alternative is to present a number of problems, and allow each pupil or group of pupils a choice. (If pupils who have not done problem solving before want advice on which problem to choose, the teacher who knows them well can give it.)

Pupils or groups who have made their own choice of problem feel more responsible for their work and more involved. A choice also allows pupils to exercise their own preferences and tastes. Their feelings of involvement and responsibility reach a maximum if they are tackling a problem which they themselves invented.

Sometimes their choice will be inappropriate, or it will be appropriate and they make no progress, or the pupil will decide that he or she dislikes a problem which had seemed initially attractive. In such cases, the problem may be discarded and another choice made.

There must also be, of course, a realistic negative side to this relative freedom of choice. When a pupil chooses a problem, that pupil has a responsibility to take it seriously. If they should decide that it is not after all appropriate, then that is a serious decision, not a flippant one. If they choose to persevere and find that it is harder than expected, the responsibility and the credit for their perseverance and achievement is also theirs, with the proviso that if the teacher is sure that the problem is way beyond them, but nevertheless leaves the pupil to flounder, then the responsibility may lie more with the teacher.

The freedom and responsibility that the pupils are exercising must be relative, because of the teacher's conception of where the pupils are going, which problably includes a syllabus and assessment, and because of their relative ignorance of mathematics and problem solving as an activity. When they are more experienced, they will know, from their experience, what the possibilities are, and their freedom and responsibility will be correspondingly greater.

As usual, they will learn these lessons more quickly if the teacher discusses these ideas with them, explicitly, in class. The more pupils appreciate the process of problem solving, the process of learning mathematics, and indeed of the process of schooling, the more easily they will be able to solve problems.

Varying the problems

No problem is a given fixed quantity. Any problem can be presented in many different ways, and the modes of presentation will affect not only the appearance of difficulty, but the actual difficulty of the problem.

Here is an illustration which, though striking, is much more than a caricature:

Version 1. The interior of a simple plane curve is divided into a finite number of simply-connected regions, each of which is assigned to one of the classes $R_i (i = 1, \ldots, n)$. If no two regions bounded by the same edge are assigned to the same class, and if the number of classes R_i is a minimum, find the maximum value of n.

Version 2. What is the smallest number of colours needed to colour a plane map, if any two adjacent regions are a different colour?

The first version uses many mathematical terms and abstract ideas, and sounds difficult for that reason alone. The second version uses everyday terms, apart possibly from 'plane', and could be understood by almost anyone, mathematician or no. It is about maps and an obvious property of maps. 'Map' is here an effective metaphor for a region divided into a set of regions. Even more powerfully, the verb 'to colour' is used as a metaphor for 'assign to a class'. Significantly, mathematicians themselves use version 2, rather than anything resembling version 1.

It is an important point that the metaphors of 'map' and 'colour' are used not merely as decoration, but to embed the problem in a real-life context that actually makes it easier to understand. (It is an historical irony that no such transformation was actually necessary – Kempe originally posed the problem to himself while colouring a map of England.)

Whether an everyday context, or a context taken from one of the sciences for example, makes a problem easier to solve, or is irrelevant decoration, or actually hinders its understanding, depends on the pupil.

Consider these two versions:

Version 1. The sum of two numbers is 50; their difference is 14. What are the numbers?

Version 2. Mr Talgarth spends 50p on a paper and a bar of chocolate. If the chocolate cost 14p more than the paper, what did each of them cost?

Some pupils will find the second version wordier and no easier than the first. Others may find the context in the second very helpful. Yet others may see some point in doing the problem in the second version, though cognitively it is no easier. The context makes it emotionally more accessible. In contrast, some pupils will prefer the first version for the same reason. It is pure and simple and straightforward.

The language in which a problem is expressed has a great effect in defining its apparent and real difficulty. A problem simply expressed may be accessible to primary school pupils, but the same problem expressed in more abstract language may be inaccessible to most secondary pupils.

Generally speaking, it is desirable that problems should be accessible to as many pupils as possible. However, because pupils are usually very sensitive to the language used, more difficult language can sometimes be used to indicate as it were, that the problem itself is difficult.

Here is an example. Which of these versions of the same problem would be most appropriate for different pupils? In particular, does the easy version (3) misleadingly disguise the fact that this is really a difficult problem, or does the hard version 1 falsely suggest that you need to understand at least some algebra to solve this problem?

Version 1. Prove that $n^3 - n$ is always divisible by 3.

Version 2. Prove that the difference between an integer and its cube is always divisible by 3.

Version 3. Why is the difference between a whole number and its cube always a multiple of 3?

The problems in these books are generally expressed in very simple language. This is appropriate for most pupils, but stronger and older pupils may translate some problems into more advanced language, and of course use more advanced language in thinking about and solving them.

Changing expectations

Pupils' expectations

Every way of working involves its own expectations, often unstated. A new approach requires new expectations. To avoid leaving pupils confused and anxious, teachers need to discuss these new expectations, contrasting them with the old expectations, and explaining the reasoning behind the changes, so that they are seen as rational rather than arbitrary.

Pupils who have been taught traditionally will need to change their expectations most. There is a danger here that a few pupils will interpret, for example, a choice of problems or the freedom to talk to their neighbours, as a sign that they can do what they like. Such conclusions are quite false. All such freedoms impose obligations which are extremely onerous and not to be taken lightly, but this may only become apparent to pupils as a result of discussion and their experience of problem solving. Similarly, other pupils may feel anxious at suddenly being offered so many choices rather than being told what to do. They can also be reassured by discussion, and the knowledge that the teacher is still there, but in a different role, which is also no less demanding than the traditional one.

Here are some expectations that pupils should learn to take for granted while solving problems, which may well clash with their previous experience.

Pupils have a choice of problems

This will not always be so, but as far as possible pupils should have a choice and should be encouraged to make their choice responsibly. Flitting from one problem to another without persevering with any one problem is as enfeebling as always expecting to be told which problem to tackle by the teacher.

Some problems are more enjoyable than others

Pupils always have their own preferences but probably do not expect them to be taken into account. Often they cannot be taken into account, but it is important that pupils should learn to recognise their own preferences, and to act both on them and on occasion against them, when this is appropriate.

By considering what they like and why, and observing what ideas are successful in actually solving problems, they will be developing their own personal aesthetic sense and providing themselves with an important tool for effective problem solving.

By recognising the idiosyncratic element in their own preferences, they will realise that their own choices are not a sufficient criterion for the value of problems or syllabus areas which are worth investigating.

24273

The teacher will not explain the method of solution

Pupils' success at problem solving does not depend on any ability to follow the teacher's explanation of a solution method. When solving problems, the pupil must decide the method to be used. If the problems used initially have been well matched to the pupils, then they will not find this frightening, though a few of them may need encouragement to strike out on their own.

Working with other pupils is good, and not cheating

The pupils' expectations must, of course, be that every member of the group will contribute, and be allowed to contribute, and the results of their joint effort and its assessment will reflect on the whole group. Given such good will and the ability to realise it, a group is often much greater than the sum of its parts.

Pupils learn from, and teach, each other

Traditionally pupils have expected to learn from their teacher. When solving problems, even if they are working individually, they can expect to learn from other pupils', possibly very different, solutions to the same problem, or to similar problems.

If they are working in a group, then they should be learning from and instructing each other all the time. This requires a receptive and modest attitude which some pupils are loth to adopt, or cannot easily adopt!

The pupils' own ideas are extremely valuable

The value of pupils' ideas is not judged by the extent to which they fit what the teacher is saying. Pupils who have confidence in themselves will value their own ideas, both the good and the bad, because they will know from experience that you cannot tell one from the other at the moment you think of them, and many ideas which turn out to be wrong are nevertheless pointers to more effective ideas.

The process of problem solving, for pupils as much as for professionals, is an amalgam of success and failure, genuine progress and detours down dead ends, fanciful notions that do not work and effective insights. If maximum use is to be made of ideas, they must be valued for themselves, not because they fit the expectations of a teacher looking over the pupil's shoulder.

Pupils' questions are themselves problems

Problems do not come essentially from books, but from out of people's heads, including pupils'. A problem might be described as a question that is taken seriously. Naturally pupils' questions should be taken seriously.

Some may be simple, some may be answered immediately by the poser or another pupil. Others will be difficult and tricky, and may well be displayed, or added to the collection of problems the pupils are tackling. The pupils themselves should be a valuable source of problems.

There is more than one way to solve a problem

Some pupils may initially be tempted to look for the one way to solve a problem that they imagine the teacher might have taught them if it was that kind of lesson. Such pupils are treating the problem as an exercise in 'reading the teacher's mind'.

Real problems are not like that at all. A problem can typically be solved in many different ways, and pupils are likely to differ in their approaches and their solutions. By comparing methods and solutions, pupils learn more about the problem, appreciate the richness of mathematics and learn to appreciate each others' abilities also.

There is no rush!

Unfortunately this statement is often false! There are usually bells at the end of lessons, a syllabus to cover and maybe examinations looming. These constraints should be discussed with pupils in the context of their own interest in persevering with problems, and the pressures on professional mathematicians.

Applied mathematicians often have to work to deadlines, and they are often looking for the best solution in the circumstances. Pure mathematicians can more often afford the luxury of working on a problem not just over several days, but over months or even years.

In contrast, pupils have traditionally been expected to finish their exercise within a short set time limit, and inability to complete every item has been a measure of their failure. That attitude and expectation, at least, is quite alien to problem solving.

Pupils will often get stuck

It cannot be emphasised too strongly that getting stuck while solving a problem is an essential part of the process. If a problem can be solved without ever getting stuck, it is no problem but an exercise. The solution of any problem naturally oscillates between apparent movement forward, which may prove illusory, and feeling stuck and wondering what to do or try next.

Pupils will sometimes fail completely

Complete failure is also an essential occasional aspect of problem solving, though if a pupil fails more than occasionally then something is seriously wrong. The teacher could be at fault in providing problems that are too difficult or allowing the pupil to tackle a problem, perhaps thought up by the pupil, which is much harder than it seems.

Pupils generally have a remarkable ability, aided by the freedom to try a problem and reject it if they feel it is too difficult, to choose problems which are within their reach, like small children who do not fall off the climbing frame as often as an apprehensive adult might expect. This ability to choose suitable problems increases with their experience of problem solving.

A few pupils, however, for psychological or personality reasons, may tend to persistently overstretch themselves, just as others may try to insist that problems which are within their reach, are in fact beyond them.

You cannot solve every problem – and neither can the teacher!

Nor can professionals. This is important, because questions asked by pupils themselves may well be insoluble by them, and maybe by the teacher. It is useful to have some unsolved problems on hand to illustrate the challenges that even professionals have not yet met – for example, Goldbach's conjecture that every even number greater than 4 is the sum of two prime numbers.

There is luck in problem solving

No amount of experience, and no amount of insight will guarantee that the method by which you choose to approach a problem will be successful. You cannot tell in advance whether a neat idea is a step in the right direction or down a blind alley. This is true, as always, from the efforts of the tyro to the work of the grandmasters.

Consequently when several pupils take widely different times to solve the same problem, or when some succeed and others fail, it cannot be concluded that the faster or more successful solvers are 'cleverer' or 'better' than their fellows.

Unfortunately pupils are quite likely to draw just such conclusions, so this important idea needs to be introduced early on, as a prophylactic against premature ranking of each other's efforts.

In time, pupils will develop their own assessment of the abilities of other pupils, independently of the teacher's assessments, based on their experience of working with each other. However, before this development appears, they will have long since learnt to appreciate the effect of luck and to discount particular successes or failures.

Success is to the pupil's, or pupils', credit

Pupils have far more reason to feel satisfied with themselves and to value their own work when they are solving problems they have not seen before, than when they are attempting to follow the teacher's exposition. And indeed, most pupils do show great pleasure in their own achievements.

This is not to devalue the teacher's role – far from it. But the teacher's role is different and more subtle, and in particular as I shall suggest below, should focus more on debriefing pupils on their experience rather than telling them what to do initially. It therefore seems less prominent.

Explaining what you have done is a problem in itself

Most problems are effectively three problems in one: understanding the problem, solving it, and then explaining the solution. The first stage usually needs only a little language, if only because many of the best problems are expressed in few words.

The middle stage often depends on images, symbols, and a kind of language that would make little sense if transcribed from a tape-recorder and which anyway may be mostly internal. This is fortunate for many mathematicians whose command of language leaves much to be desired.

The final stage, for such mathematicians, usually involves little language either, no more than a code to string the mathematical argument together. Pupils have no experience of any such code and their efforts to explain clearly and lucidly what they have done and why their conclusions should be believed depend heavily on their fluency in written language.

Because there is only a modest correlation between mathematical and linguistic ability, it follows that some pupils will solve problems very well, but have serious difficulty in explaining their solutions, while others, perhaps less powerful mathematically, will be able to explain themselves beautifully.

Insisting that a pupil work hard on expressing their ideas clearly is one way to develop the pupil's expressive ability, though it may be far less enjoyable than actually solving a problem.

In the meantime, the variety of ability needs to be recognised and credit given where it is due, for the mathematical and the linguistic achievement separately, not least because both aspects are separately important in real life where most mathematicians are not expected to be linguistic stylists, and there are many jobs which demand a modest level of mathematical ability together with linguistic fluency.

Solutions are not simply right or wrong

A pupil's efforts will be judged as a whole, and not simply by checking to see whether 'the answer' is correct. Even if the problem has a single answer, the method by which it is found, the ideas displayed, the insight or ingenuity shown, are all worthy of credit. If the problem has no single answer, then there are even more factors to take into account. The teacher might appropriately respond with a written comment, plus a mark if that is considered necessary which often it will not be.

Teacher's expectations

The teacher also has to adapt to new expectations. Here are some suggestions.

Be prepared to handle the clash between the pupils' old and new expectations

Problem solving may be a new experience for many pupils, who are likely to respond positively and favourably. It does not follow that they will get the most out of problem solving if they are left to make sense of this approach by themselves.

What happens when they fail? How did John get the solution in a couple of minutes after Peter and Mary had taken hours? 'I know how to do it but I can't explain it!' These are serious problems, which deserve to be treated seriously.

Solve problems yourself, including in the classroom

As long as you are a traditional source of authority, you can afford to rest on your laurels. After all, you passed your examinations years ago! Problem solving isn't like that. It is something to be done here and now, and something to go on doing.

Your own problem solving will improve, and pupils will take you as a strong role model if you solve problems, and they see you solving problems, or trying to. In particular, you can show your appreciation of their ideas by taking some of their more difficult questions as problems, and try to answer them as problems.

Look at problems from the pupils' point of view

You will be able to solve many problems easily by applying your book knowledge, which the pupils do not possess. It is therefore very important to practice seeing problems from the pupil's less knowledgeable point of view. One way to do this is to make a point of *not* solving problems by the methods you know, but always looking for different and less usual methods.

At least as important, and marvellously instructive, is to listen to pupils and observe them, in order to try to get under the skin of their thinking.

Recognise and expect to learn from pupils' ideas

All pupils, from the youngest to the oldest, think of ideas, make suggestions, and ask questions, which are valuable in themselves, and can be enjoyed and appreciated by you without dishonesty and without patronising them.

The more confidence they develop, and the more they trust you, the more readily they will make comments, knowing that their ideas will not be dismissed as irrelevant or silly.

Encourage discussion and argument and emphasise the importance of language

Many pupils will continue to turn to you as a source of authority and a supplier of answers to questions which they can answer themselves, or in discussion with other pupils. They may often turn to you before they have given the matter any serious thought themselves. Simply turning the question round and asking them what they think will force them to rely on themselves first.

When pupils discuss and argue among themselves they are testing their mathematical ideas, and also developing their use of language, and their ability to think logically and rationally. These abilities are closely intertwined, though by no means identical, and developing any of them in a mathematical context is likely to develop the pupil's overall mathematical ability.

Do not give away solutions!

When a pupil genuinely needs help, and the alternative is anxiety and unnecessary failure, it is easy to give away more than necessary. It is desirable to give away as little as possible, not least in order to avoid the implication that the pupil is nowhere near the right track. One of the teacher's greatest challenges in problem solving is to find the right words that will set a pupil off again without undermining the pupil's own achievement.

Notice new problems, and make sure they are recorded for future use

Pupils will ask questions and suggest problems and forget them immediately. It is in your interest as much as theirs to make sure they are not forgotten, both so that pupils know that their ideas are valued, and so that you build up a collection of problems for future use.

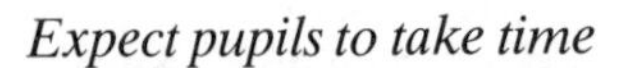

Expect pupils to take time

It is often claimed, especially of apparently weaker pupils, that they cannot concentrate for more than a few minutes at a time.

The activity of problem solving more often than not gives the lie to such claims and reveals that in the right circumstances they are willing and able to focus their attention for long periods of time on problems in which they are involved.

Similarly, in answering questions in response to the teacher's exposition, pupils are expected to answer quickly while the rest of the class is waiting. Such situations are extremely artificial. When pupils are solving problems, questions can be mulled over, rather than responded to immediately under pressure.

Since you are their role model, you can demonstrate this possibility and the seriousness with which you take their questions, by taking time to answer their questions, and accepting a question, when appropriate, by saying that you will think about it and give your answer later.

Accept HOW mathematics is difficult

Mathematics is difficult absolutely, in the sense that the greatest mathematicians have not mastered it. There have always been problems they could not solve and the advance of mathematics has produced more problems rather than fewer.

It is also difficult in quite a different sense. Mathematicians could be satisfied spending their time tackling easy problems which they were confident they could solve. In this way they would experience continued success at a modest level. In fact, of course, they do nothing of the sort. Like mountaineers who are not satisfied to climb only the smaller and easier peaks, professionals aim at the hardest problems they feel they can manage.

They deliberately make things difficult for themselves, because in overcoming greater obstacles they feel far greater satisfaction, and also receive greater credit from their peers.

Pupils also should expect to tackle problems which will stretch them, and not be satisfied with problems which are too easily within their reach. Fortunately, most pupils are very willing to challenge themselves, and do not have to be persuaded. However, it is important that they appreciate the element of personal choice here and that they examine their own feelings and decide at what level they will take aim.

Accept that explaining, recording and communicating a solution is a problem in itself

This point has already been discussed. It is important for pupils who have difficulty in explaining their ideas that every problem enjoyably solved should not have tagged onto the end an activity that is stressful and unenjoyable. Pupils who are poor at representing their ideas and arguments can appreciate that explanation and recording of ideas have an important role, and learn bit by bit to improve their performance in this direction, without feeling that no solution is acceptable without a complete write-up.

It is notable that professional mathematicians do not write up everything they discover – to do so would be absurdly time-consuming. They keep notes of important ideas, and write up results for particular reasons, such as publication. Pupils can learn to do the same, and value all the more the work that is written up completely.

Expect most pupils to improve at problem solving!

Traditionally, pupils have had two difficulties in learning mathematics: the difficulty of the mathematics itself; and the difficulty of learning through someone else's exposition, which is one of the hardest ways of learning anything. Learning through problem solving, they should be almost entirely released from the second difficulty.

Add to this release the challenge and excitement of problem solving and the effect is to improve the mathematics of a large majority of pupils, especially the many pupils in lower sets who are there as much because they are especially weak at learning through exposition as through any lack of mathematical ability. This is not to say that they will suddenly all appear to be brilliant mathematicians. They will not. It is to say that they will get much nearer to their maximum potential as mathematicians through a problem-solving approach.

The other side of this coin is that a small number of pupils, mostly in the higher sets, are there because their ability to learn from exposition is greater than their mathematical ability. The performance of such pupils may drop relative to their peers. Fortunately, they tend to be excellent at writing accounts of their work, credit for which will help to disguise an otherwise distressing change in their fortunes.

Working in groups versus individual work

An increased emphasis on problem solving in recent years has been paralleled by a greater emphasis on group work. This creates the dangerous trap for teachers of supposing that problem solving implies working in groups.

It does not. Each mode of working has its own advantages and disadvantages, and a fair summary might be to say that all pupils should be capable of and have experience of both modes, although individual students will also have a personal preference for one mode or the other. Even this is an oversimplification, because working with a partner or a couple of friends, for example, is very different both from individual work and working in a group of half a dozen other pupils whom you do not know so well.

Starting up

A class of 30 pupils each of whom is working individually and separately is not helpful to you, or the pupils. This is especially so when pupils start problem solving.

Some of them will lack the confidence that comes from experience, and will be inclined to fall back on the tired old assumption that the moment they get stuck they must call for your assistance. In time they will learn better, but they cannot reasonably change their expectations overnight.

Pupils in a group can share their difficulties, through invaluable talking and discussion and if they do finally admit defeat, the defeat is shared rather than individual.

Discussion between you and your pupils is also facilitated by grouping. Thirty pupils will get, on average, less than 2 minutes of your time, each, in 1 hour lesson. Eight separate groups can expect 5, 6 or even 7 minutes of your time.

The social advantages of cooperative work

In an effective group, members will work together on a problem, discussing, arguing and agreeing or disagreeing with suggestions from individual members, who will stand up for their own ideas but also be prepared to listen to others and to abandon their own ideas if they do not survive examination.

Routine tasks, such as making many calculations and entering them in a table, will be organised fairly within the group, and completed more quickly.

If an account of their activity is to be written up in detail, the work will be allocated fairly among the members, and none will either escape or be weighed down with an unfair share.

All the members will feel part of the group and have an added confidence because of their membership, and the work of the whole group will be much greater than the sum of its parts.

The cognitive advantages of group work: discussion and argument

The cognitive advantages of group work can be just as great as the social. One pupil can only bring to a problem his or her own understanding and ideas. This can be a severe limitation. For example, it is easy for one solver to get wrapped up in one idea, which could be a dead end. Even if the solver realises that a new idea is needed, it may prove difficult to escape from the original perspective.

In an effective group, ideas are tossed around, different members have different points of view, and it is clear that there is more than one way to examine the problem.

In group discussion, ideas are examined, explored and clarified, and pupils develop their command of language. In the short term, language is needed to clarify the meaning of the problem, which may well be intentionally or unintentionally ambiguous, to discover its solution and maybe to record the group's deliberations.

By using language under modest but real pressure pupils improve their ability to converse, discuss, argue and explain.

Demonstrating clearly why the sequence 3, 6, 12, 24, 48... can never include a perfect square is a problem in mathematics-plus-English, even when the solver 'sees' a solution clearly, which requires the careful organisation of several ideas, and their expression in language which makes the ideas and their relationship clear to the listener or reader.

The problem of dominant individuals

The comments so far have referred to an 'effective' group. Unfortunately, most groups are more or less ineffective. Simply placing pupils round a table with a common problem to solve does not ensure that they will enjoy the benefits of effective group work.

It is more likely that a few confident individuals will dominate the group, imposing their ideas on the others who are organised to carry them out. The less confident members, who may well be the mathematically weaker members, do not have the courage to argue and end up as the dogsbodies, contributing little or nothing.

Discussion takes place almost entirely between the dominant members who only defer, if at all, to each other, and who are quite probably in happy mutual agreement much of the time.

Such a group may give an impression of activity, and indeed may be very productive in its own way, but it has none of the advantages of the good group. The members would almost certainly gain if the dominant members were packed off to do their own thing by themselves, giving the other members more opportunities to speak, more chance to exchange ideas, more respect for their ideas and more to think about.

Dominant individuals are already recognised as a problem in mathematics classes in another respect. Boys tend to dominate girls, by monopolising the teacher's attention at the expense of girls, by interrupting and shouting out more, but not giving way when girls want to respond to what the boys say, and by demanding first rights to resources and equipment. Recent research suggests that this factor is so strong that even teachers who are aware of it, and make a conscious deliberate attempt to counter it, only partially succeed. There are social and psychological reasons for this behaviour. Girls are still brought up to be relatively modest, shy and retiring in the presence of men, while boys are expected to be boisterous, to interrupt and hold the floor, as men do in the presence of women. Psychologically, boys tend to overestimate their own abilities, and their social charms, and to be unreasonably confident, while girls underestimate their abilities and unreasonably lack confidence. Boys tend to perform in examinations below their own expectations, but girls perform above theirs.

Problem solving, in mathematics and other subjects, at least as much as traditional class teaching, can give boys ample opportunity to be boringly cocky, overconfident and obnoxious. It may even offer greater rewards. Many an explorer has reached an apparently absurd goal by obstinate bloody-minded determination. Giving up quickly through initial lack of confidence, in contast, guarantees failure.

In the long run, dominant pupils have the opportunity to learn to respect others more, to learn from their ideas and to appreciate the benefits of cooperation. In the short term, their effects are often disastrous.

The advantages of individual work

Working in groups has no monopoly of virtues. This is fortunate, because some pupils much dislike working with others and will happily and successfully work in isolation or perhaps with a single friend. (Pupils who work alone because they are rejected by others are a different matter.) They enjoy the advantages of doing all the thinking themselves and taking all the credit; at least this is how they may see themselves, though in practice most apparently isolated pupils are in touch with others, checking on their progress, and swopping ideas, (and sometimes jealously refusing to reveal anything at all.)

More or less isolated pupils avoid the hassles of disagreeing with others, or feeling obliged to go along with the majority when they would prefer to follow their own ideas, which they are free to pursue with complete concentration, and which may indeed be excellent. The advantages to their own thinking can be considerable; it is a notable fact that few professional mathematicians publish joint papers.

The disadvantages, which can be severe, are lack of practice in communicating with others, and some loss of language use in general, though it is an important point that the individual's thinking often involves internal language use.

A variety of environments

There is no doubt that all pupils should have experience of working individually, in small groups and in larger groups, including partaking confidently in whole-class discussions, which are often typical of poor groups is actually involving only a proportion of the whole class.

The best environment for a particular pupil most of the time is a moot point. The differences between pupils who prefer to work alone and those who prefer to work with their peers seem largely psychological. Therefore, they may be modified by experience, but are unlikely to disappear completely.

This need cause no problems of organisation if the pupils who choose to work alone accept that they will receive less help in proportion because what help they do receive is focussed entirely on one individual. Since the chances are that they do not want the teacher to interfere with their work either, this is likely to be a fair bargain.

As so often, the answer to the complexity of individual preferences and cognitive and social styles is to combine a variety of ways of working, rather than oversimplifying and heading for either extreme.

Mathematical problem solving

Heuristics

Anyone who has solved many problems will appreciate that the same general plans of attack turn up repeatedly. Some problems may be reminiscent of an already known problem; some are much clearer when we have drawn a picture and can see what is happening; some break down naturally into several stages that can be tackled one by one.

Such principles are so general that they are not limited to mathematics, but apply to all problem solving. Indeed, they are familiar in everyday life. If the car breaks down, then we may recognise the symptoms and at once suspect the fault, or we may consult the car manual to see more clearly what might be happening, or we may unscrew the 'doings' to get at the 'thingummybob', solving the problem one step at a time.

Because these principles are so general, there is a danger that too much will be made of them. Pupils who take naturally to problem solving will use heuristic principles without even thinking about what they are doing, and stopping to make them think may simply confuse them. Try describing to yourself how you walk downstairs as you do so and you may fall over!

On the other hand, pupils may get stuck on a problem, especially when they are inexperienced, because they have not thought of looking at a particular case, or doing an experiment, or working backwards from the desired goal. Such pupils can be set on a profitable course by suggesting a suitable heuristic.

Very weak pupils may see the point of different heuristics only with difficulty, and will have most need of explicit discussion of their use.

To summarise in a rough-and-ready rule: the stronger the pupils, the less the value of explicit discussion of heuristics; the weaker the pupils, the more necessary is such discussion. Ideally, the use of heuristics should be second nature, taken for granted rather than the subject of a song and dance.

Since the teacher has to deal with all sorts and conditions of pupils, the teacher should have an explit understanding of heuristics. One way to develop this is to solve problems yourself and afterwards analyse the process by which you solved them. This will also give you insight into your own idiosyncratic biases in problem solving! Studying pupils' solutions will show how they use, or fail to use, heuristic principles.

Another possibility is to study published analyses of heuristics of which the most famous is George Polya's classic *How to Solve It,* which is now republished in paperback as a cheap Open University set book.

Here are some of the commonest heuristics.

Compare with a previously solved problem

It is difficult to find any problem which does not resemble a know problem, and the resemblance often suggests a promising approach. Naturally the greater the solver's experience, the greater the chance that a similar problem will spring to mind.

This heuristic is related to looking for analogies and metaphors.

Draw a diagram or make a model

Geometrical relationships, patterns in numbers, rates of change, are often more clearly perceived, if only informally, through the sense of sight. Of course the diagram or model must be appropriate.

Make a model

Models do more than appeal to the sense of sight. The best models can be turned round to provide viewpoints, literally, that would be very hard to visualise mentally (the skeleton of a cube) or they can be taken apart to reveal the relationships between the parts (a tetrahedron dissected into tetrahedra and an octahedron) or they can be manipulated to reveal new possibilities (a parallelogram dissected into a square) or they can be used for physical experiments.

Invent your own notation

A good notation has some of the effects of a good picture. It focusses on what is important, aids memory, and actually makes thinking much easier. Pupils are naturally expected to learn the standard notations which everyone uses, but there is ample scope for them to invent their own notations also, just as professional mathematicians do when it seems appropriate.

Break the problem into two or more stages

Beginners easily make the mistake of thinking that there must be some brilliant idea that would solve the problem at a stroke, which they have failed to spot. This is almost always false! It is more effective to look for some step you can take, any step, that will move you forward, to be followed by another step, and so on.

The first step might be to apply one of these heuristics, by calculating some data, doing an experiment, making a model, checking some special cases, anything that will possibly provide more information and greater understanding.

Guess

If you can guess the answer to a problem, you are in as strong a position for proving the answer as if it had been given to you initially. Even if you cannot guess an answer, you may be able to guess at a relationship or pattern. Working on your guess will deepen your understanding of the problem, even if the guess turns out to be wrong.

Guessing is related to experiment and pattern spotting.

Work backwards from the solution

In many problems you either know or strongly suspect the solution. The problem might be to prove that such-and-such is indeed so. An experiment may have revealed the solution. You may have made a convincing guess. It is often easier to work backwards from the solution than to work forwards from your starting position. Compare visiting a strange address and returning home afterwards!

Do an experiment

Archimedes discovered the positions of the centres of gravity of complex solids such as paraboloids of revolution by experiment, and then used his experimental knowledge to prove his conclusions mathematically.

Pupils can also do physical experiments, for example to find areas or to discover the effects of forces represented by vectors.

However, many more mathematical experiments require only pencil and paper or a calculator – or a computer. The data from the experiment may then point to the solution, often by showing a striking pattern.

Spot a pattern

This is the twin of 'Do an experiment'. To spot a pattern you need some data, and to make sense of data you need to spot a pattern.

Vary the conditions of the problem

By studying closely related but slightly different problems you will often get insight into the original problem. If you are lucky, you may realise that it is one case from a set of problems which are easier to understand as a set than individually.

If you vary the conditions of the problem systematically, then you can reasonably expect to find a pattern in the results.

Generalise

This is another way of expressing the previous heuristic. Seeing one problem as representative of an entire class of problems often makes the original problem easier, rather than harder, to solve.

Study special cases

Special cases are often very revealing. For example, if the problem is about quadrilaterals, it may be instructive to consider what happens if the quadrilateral is a square.

Special cases are useful in testing guesses or conjectures, not least because it is often easier to test a special case. Do you have a conjecture about prime numbers? Does it work for 2, the only even prime? Do you have a conjecture about circles? Does it work when the circle degenerates into a point, or expands into a straight line?

Transform the problem into another problem

Descartes showed how problems in pure geometry could be transformed into problems in algebra by the use of coordinates. Transformations from geometry into arithmetic or algebra, or vice versa, are among the commonest heuristics used by mathematicians.

It is typical of modern abstract mathematics to focus on the essential features of a problem, distinguishing them sharply from the less essential. A natural consequence of this abstract approach is to reveal that many problems that seem superficially to be different are essentially the same, and can be transformed into each other.

Look for analogies and metaphors

This is closely related to the last heuristic, but the terms 'analogy' and 'metaphor' emphasise the generality of these heuristics. Both terms are familiar not only to scientists but to poets and novelists, painters and architects. It is no exaggeration to say that much of our thinking is metaphorical and analogical, and that many of these heuristics simply emphasise the importance of looking at the same thing in different and perhaps more illuminating ways, or finding a viewpoint from which apparently different objects appear to be related after all.

Use proof as a heuristic

Proof should not be thought of as only a final step, a stringing together of the entire solution in a logical sequence which will convince not only the solver but also a critic.

Try to find reasons for, try to prove, your guesses and conjectures, the patterns you have spotted, the result of your experiment, and you will learn more about the problem. This is as much as to say, 'Be critical! Be severe with yourself! Don't be satisfied to examine only the surface, dig deeper to find out what is really going on!'

General concepts

Below the heuristic principles common to every problem-solving activity in every walk of life is another level of general ideas more specific to mathematics; below this are yet further levels specific to parts of mathematics only, to particular topics or types of problems (though there is no sharp dividing line between the levels).

These general concepts cannot be classified as content, since they are only fallible (though extraordinarily useful) guides and pointers, like the heuristic principles already discussed, rather than true theorems. Consequently they do not appear in textbooks, but they are taken for granted by experienced problem solvers and are indeed one of the major tools that the solver brings to a problem. They represent in words some, only, of the feelings, intuitions, expectations, proverbial wisdom and seasoned judgement of the mathematician. The better the mathematician, the more powerful will be his or her general concepts.

Because they are so specific, they cannot be described as common sense, and pupils will rightly not take them for granted since they depend on a broad experience of mathematics which pupils, at least initially, will not possess. Fortunately, they are usually much easier to understand than genuine mathematical theorems, and it is not difficult for pupils to appreciate them. They are therefore excellent subjects for comment and discussion, in the context of pupils' actual problem solving.

Here is a handful of the general concepts that seem to me of most value to pupils. They are in no particular order.

If you put a pattern in, you will get a pattern out

If you calculate a function of 1,2,3,4,5,..., you can expect a pattern in the results, and likewise if you calculate the same function of 1,2,4,8,16..., but picking the arguments 2,17,5,83 and 55 will not produce a simple pattern unless there happens to be a special relationship between the function and these numbers.

If you construct a pattern according to some rule, for example a sequence of numbers or a tessellation, then there will certainly be other, different, patterns in the same sequence or tessellation. In other words, every pattern contains many patterns.

Mathematics is beautiful. You can expect patterns everywhere in mathematics

It is difficult to say exactly what a pattern is, and harder to say what makes mathematics beautiful, but there is no doubt that mathematicians expect beautiful patterns to appear and that this expectation is a valuable pointer to success.

If pupils are to develop their mathematical ability to the full, they should be encouraged to make aesthetic judgements on their own work and on the work of others, while recognising, as the teacher must recognise, that such judgements have a considerable idiosyncratic element.

Look for symmetry everywhere

This is another version of the last idea. Symmetry need not be, and in advanced mathematics seldom is, simply visual. The coefficients of a quadratic equation, for example, are symmetrical with respect to the roots. This symmetry can appear visually, after a fashion, on a graph of the function, but it is certainly not apparent in the usual form of a quadratic equation.

Mathematicians naturally search for and expect to find such hidden symmetries throughout mathematics.

Many problems cannot be solved, and you can prove this

Pupils often find this idea surprising, and yet it can be an excellent introduction to ideas of proof, because it is often more natural to ask of something which doesn't work, 'Why not?' than to ask of something which works perfectly well, 'Why?' This, after all, is what people do in everyday life – take the world for granted until something goes wrong, and then start wondering and asking questions!

Funny things happen at infinity. Infinity is weird

Many of the difficulties of mathematics, historically speaking, have arisen through attempts to get to grips with infinity. Pupils can experience and enjoy some of this weirdness.

Can an adding sum go on for ever, and yet the total never exceed 1? What happens when a circle gets bigger and bigger? Does it become a straight line? Does the sum of an infinite number of fractions have to be another fraction?

0 and 1 are special

You cannot divide by 0, so 0 has no reciprocal. A number does not change when zero is added or when it is multiplied by 1. Unity is its own reciprocal and all its powers are unity. Zero and unity are rather like odd and even. Early mathematicians from the Greeks onwards made observations like these, which are much more than curiosities; they are pointers to the profound role of these numbers in more advanced mathematics.

Some problems can be solved by algorithms, some cannot

The important point for pupils to appreciate here is not the nature of an algorithm, though this is vital and ought to be a part of the context of any mathematics course, but the striking difference between problems which can be solved by straightforward methods, and those which depend on insight and imagination.

In practice, most problems require a mixture of both. Pupils will be better solvers if they appreciate what they can do with little or no insight, and when a leap in the dark is called for.

Interesting things happen at extremes

In particular, extreme situations in mathematics often have rather simple features. This makes them well worth investigating. The maximum area is enclosed by a loop of string when it is a simple circle. The product of two numbers whose sum is constant is a maximum when they are equal.

The special features of 0 and 1, and of infinity can be interpreted as special cases of extremes.

Even when everything seems to change, something stays the same

This is a way of saying, without using technical terms, that every transformation has its invariants. The invariants of the transformation, if you can find them, are a kind of fixed point, something to get your bearings by. They are simple, as a line of symmetry is simple.

Metaphors and analogies in mathematics can be exact

The idea that there are as many numbers in the sequence 1,4,9,16...as there are in the sequence 1,2,3,4...can be more than a vague idea; it can be expressed exactly. If you think that a cube and an octahedron are similar in a way, well, that is so, but you can express it more precisely than that (and in several different forms).

This idea naturally promotes an emphasis on thinking about what you are saying and using language as carefully as possible.

Mathematical objects can be looked at in different ways

A cube is not just six squares joined together. It is also a framework of edges, the join of eight symmetrically positioned points, and many other things.

Mathematical language can be interpreted in different ways

A fraction can be thought of as a number, and a ratio of two numbers. A straight line may be a Euclidean straight line, or it could be the shortest distance between two points.

Such variations emphasise the extent to which mathematicians are continually thinking metaphorically.

If a problem has several solutions, there will be relationships between them

A positive number has two square roots. Their sum is always zero. The solutions to a 3×3 magic square using the numbers 1–9, are connected by rotation and reflection. There is a simple connection between the two triangles with two given sides and non-included angle.

This idea not only emphasises the depth of pattern in mathematics, but can be a powerful aid to finding solutions — for example, finding the missing solutions when you have one or two.

Proof and certainty are possible in mathematics

It is a moot point whether proof should have appeared at all in the list of heuristics, interpreted as the most general problem-solving principles, because the idea of mathematical proof is so different from ideas of proof in everyday life, in a law court for example.

There is certainly a concept of proof which is shared by mathematics and abstract games, say, but is not shared even with the other hard sciences, except when they are using mathematics.

The idea that mathematics offers unusual possibilities of certainty and conviction is difficult for many pupils to grasp, and needs to be discussed in the context of actual examples of simple arguments if they are to develop an awareness of proof and inference in mathematics and be able to produce simple proofs and convincing arguments themselves.

It is worth looking for analogies of features in two dimensions, in one and three dimensions

I have included this idea here because it illustrates, as does the idea of the special nature of 0 and 1, the level of more specific concepts which are not applicable even to the whole of mathematics.

This does not make them less useful. Far from it: their lesser generality is compensated by their specific applicability. Whatever field a mathematician is working in, he or she will develop a range of ideas which have a vital heuristic role. Pupils will inevitably do the same, though their ideas are usually hidden because they are no part of the syllabus content, are not assessed in examinations and are only made partially explicit in comment and discussion, a large proportion of which revolves round just such ideas. Problem solving especially promotes the development of these valuable ideas because it is such a rich activity in which pupils spend so much time thinking their own thoughts and trying to make sense of their experience.

An example with a commentary: quadratic equations

The topic of quadratic equations seems an ideal example of teaching the syllabus through problem solving, for several reasons. It is a thoroughly traditional topic. There can therefore be no suspicion that problem solving is an approach only suited to 'modern' topics.

It is very rich topic with many relationship to science and to other parts of mathematics. This richness is largely stripped away by a treatment which concentrates on the standard methods; a problem-solving approach naturally brings it out.

It tends to be thought of as a difficult and advanced topic. It easily can be, but it can also be tackled, through problem solving, at a more elementary level, which will in turn open the way towards a more abstract and deeper understanding.

There are also many aspects of quadratic equations which appear naturally through problem solving and which are important and useful, but which tend to be completely overlooked in treatments which start at a higher level. The result is so often that pupils who are strong enough to understand the traditional solutions well enough to answer examination questions actually possess a shallow and weak intuitive understanding, which does not stand them in good stead in more advanced work.

How should quadratic equations be presented in terms of problems in order to maximise the number of pupils who can appreciate the problems and tackle them successfully? A first step is to:

Remove and examine all assumptions about how the problem should be solved, when and by whom

Quadratic equations typically look like this:

$$x^2 - 7x + 12 = 0$$

or this:

$$3x^2 - 5x - 2 = 0$$

They are designed to be solved by factorisation, in which case it is implicit that the roots will be integers, or just possibly fractions, or by graphical methods, or by the use of the formula or by completing the square, in which case the roots could be mixed surds but not complex.

Different methods correspond to different levels of achievement at the old CSE or GCE. The topic of quadratic equations is assumed to be difficult and only taught to older and stronger pupils anyway.

The specific problem set is almost always to solve a specific equation. A much higher-level activity might be to find the coefficients of an equation, given its roots.

All these features of quadratic equations as a syllabus and examination topic are historical and contingent and must be discarded for the time being as inconsistent with a problem-solving approach. Eventually, for those pupils who require them, and are strong enough, these methods will reappear, but as end products of tackling the more general problem, rather than as the standard, and only allowable, approaches.

Laying these standard and powerful and advanced methods on one side, the problem can be looked at with a fresh eye, much like a mathematician meeting the problem for the first time.

Naturally, the more experience that a teacher has of meeting other topics for the first time, and the more experience of problem solving with pupils, the easier it becomes to adopt an innocent point of view.

Present the problem as simply as possible

The first equation quoted above states that the square of a number, less 7 times the number, plus 12, is zero.

This is not the only way to effectively express the same problem. Historically, it might well have been expressed as 'the square of a number, plus 12, equals 7 times the number,' avoiding all negative terms. The custom of placing all terms, some negative if necessary, on one side, the whole equated to zero, is a later development depending on familiarity with negative numbers and zero.

In whatever form the problem is conceived, it can be expressed in algebraic language in several ways, such as:

$$x^2 - 7x + 12 = 0$$

$$x^2 + 12 = 7x$$

$$N \times N + 12 = 7 \times \mathrm{N}$$

From a sufficiently advanced point of view these are all equivalent; from a teaching point of view they are not. The first, standard, expression assumes a familiarity with subtraction of algebraic terms which the others do not. Both of the first two assume a familiarity with the notation for squares which the third does not. Neither does the last assume that pupils cannot confuse $7N$ with seventy-something.

They might all seem to assume that pupils can read algebraic statements and expressions. This would certainly be taken for granted when the topic of quadratic equations is introduced to fourth- or fifth-formers. That assumption however can and should be questioned. Very many pupils can perfectly well understand the problem as expressed earlier in plain language, and can learn a great deal from tackling it, who are not familiar with 'reading' algebra. Lacking any prior introduction or understanding of algebra at all, they can easily understand the algebraic statements above provided that the latter are introduced in terms which they can grasp hold of. The metaphors of a code, or a foreign language, or a shorthand, all of which require translation into ordinary language, but all of which are not difficult to translate once you understand them, are useful here. (The foreign language might be an exception; it is also perhaps a metaphor biased slightly towards girls, just as the idea of a code may be slightly biased towards boys who meet them frequently in their comics.)

By considering both the variety of possible expressions and by interpreting the algebraic statement as a condensed way of writing a problem that is easily translated into familiar language a teacher can open up the topic of quadratic equations to pupils far younger and far weaker than usual. This possibility is not merely theoretical. There are good reasons for using such problems with younger/weaker pupils. In particular, it provides an excellent example of what I shall discuss below as 'true', rather than 'false', simplicity. I will only comment here that any problem that can be expressed in language that the pupil can easily understand, but which in its solution offers a wealth of possibilities, has the qualities not only of many of the best puzzles, but also the qualities, historically speaking, of the most productive problems tackled by professional mathematicians.

Initial questions and queries

Since it is plausible to suppose that older/stronger pupils can do anything that younger/weaker pupils can do, suppose that the problem is presented to a young/weak class initially in the form:

$$N \times N + 12 \quad = \quad 7 \times N$$

The coefficient of N^2 is unity, N^2 itself does not appear, and there are no negative terms. There are also two small positive integral solutions. The solution has been 'fixed' by the teacher. This form of presentation is as simple as it could be, but still contains potentially all the richness of quadratic equations in general.

(The isolation of the equals sign here is deliberate, to emphasise, and surreptitiously teach, the idea that it is different in kind from the operations on either side.)

In discussing such a problem, and simple variants, many questions naturally arise:

- Does this particular problem have a solution at all?
- Why does it have two solutions, at least? (If that fact has been spotted.)
- Would it have a solution, (two solutions, or more?) if the numbers 12 and 7 were changed?
- Apparently, as a result of experiment with different numbers in place of 7 and 12, the answer to the last question is sometimes yes and sometimes no. Why and when?
- How did the teacher's problem happen to have two solutions? By mere chance? If the teacher produces such a problem 'out of a hat', how is this trick done?
- Does the answer to such a problem always have to be whole number?
- Could it be a fraction? A negative number? (If pupils are familiar with negative numbers.)
- Can such a problem have 1, 3 or 4 or more solutions?
- When a quadratic equation has any solutions at all, they depend on the coefficients. How? What is the connection between, in this case, 7 and 12, and the solutions 3 and 4?
- What about changing the equation, so that, for example.

  $$N \times N = 7 \times N + 8 \quad \text{or} \quad 2N^2 + 17 = 8N$$

- For pupils unfamiliar with negative numbers, a problem of the first type has only one solution. Why?

Every one of these questions is a genuine problem in itself, worthy of investigation.

Natural developments and general concepts

When the pupils attempt to answer some of these questions, working individually or in groups, certain ideas and possibilities are sure to occur to them.

Initial guessing, trying out numbers almost at random naturally gives way to the realisation that the closer the values of the expressions on each side of the equation, the closer you are to a solution.

When a table shows that, for example, 6 makes an expression too great but 7 makes it too small, it is easy to wonder whether another number in between might not fit perfectly. Calculators are an obvious aid here (and computers!). Are there indeed solutions which are not whole numbers? How accurately can they be found?

Such experiences are important in developing a general concept of continuity in mathematics.

Keeping records of trials naturally produces, in effect, tables of the values of expressions. The patterns in such tables are quite easy to spot. Almost all children who study a sequence tend to look quickly towards the differences between the numbers. In this case the pattern of differences is very simple.

Pupils who know no algebra but who have studied simple number patterns will recognise the patterns that appear. Such experiences reinforce, or help to develop, the general concept that patterns can be expected to appear in any mathematical situation.

In particular, such tables show a striking symmetry, although there is apparently no symmetry in the original equations. Also, the symmetry is about either one value, or a pair of values. Why this variation? Does symmetry about a pair of values suggest trying half-integer values?

It is an important feature of such tables and the calculations that produce them that they help pupils to develop an intuitive feeling for change in the value of x, or N, and in the expression involving them. Although at the lowest level, as here, this letter will have started as a sign for a single, specific, missing number, its treatment in attempts to solve problems of this type inevitably edges smoothly and little by little towards a more sophisticated concept of the letter as a variable.

When pupils realise that a pair of solutions are related to the coefficients, they are experiencing a simple example of the general concept that there is a relationship between the solutions of a problem. With sufficient experience of problem solving, they will approach a problem with this expectation, and so such patterns will be easier to spot.

Representations and other heuristics

A standard heuristic device which works exceptionally well here is to draw a picture or representation. How can a picture be drawn of this problem, in order to show what is happening?

Pupils will be familiar with simple charts and graphs, at least. A simple graph of the values of the equated expressions shows vividly the points in the table where the values match.

Alternatively, since pupils will be focussing anyway on the differences between the two expressions, drawing a graph of their difference and looking to see where it is zero is a neat idea. The fact that the picture of this difference looks like the picture of the left-hand expressions might seem surprising. In fact all the pictures look remarkably similar in shape, even if the coefficient of x^2 varies.

Such visual representations suggest strongly why there are zero, one or two solutions, but never three or four. They may suggest further possibilities to pupils who are familiar with negative numbers.

Naturally, such illustrations can reveal some of the symmetry hidden in the problem, including the relationship between the coefficient of *N*, or *x*, and the axis of symmetry of the graph.

Notice, however, that they are drawn as a part of the problem-solving process, not as an exercise in response to precise and detailed instructions from the teacher on 'how to draw a graph of a quadratic function'. How to draw the picture in order to represent the data of the problem most strikingly is a part of the pupils' problem. In particular, should the graph be continuous? This will depend on pupils' ideas of continuity and their ideas about non-integral solutions. (Of course, the standard graphical picture will be strongly hinted to pupils who can use a computer and know sufficient PLOT instructions.)

Where random guessing may reveal nothing, systematic experiment soon reveals patterns and relationships, illustrating 'put a pattern in, get a pattern out'. Systematic data on one equation, perhaps suitably illustrated, will indicate the number of roots and their values, or approximate values. Data collected systematically on a set of equations will illustrate how the number of solutions and their values depend on the coefficients, independently of the simple arithmetical relationship between them.

What is the simplest possible quadratic equation? What is the difference between a quadratic equation and a much simpler equation, such as $2x+5=10$? Focussing on a special case and comparison with related but more familiar and easier problems might suggest that it is the squared term which is entirely responsible for the difference.

What is the behaviour of $N \times N$ or x^2 by itself? How does this explain the number of roots? How does it explain the symmetry that surprisingly appears?

Proof of relationships, at any level, is harder than merely spotting relationships. The latter is exceptionally important. Many of the greatest mathematicians, including notably Gauss and Euler, have been superb experimenters, and many important mathematical theorems have been 'known' as a result of experiment; they have been confidently conjectured long before they were proved. Unfortunately, not all conjectures turn out to be correct, and this is certainly true of pupils' conjectures.

There are many opportunities in a problem such as this for older or stronger pupils to prove their results. However, there are also opportunities for younger and weaker pupils to do the same, provided 'proof' is understood to include any secure inference, however simple, in contrast to the long and complex deductions that have, only, traditionally been labelled 'proofs' in school mathematics.

It is important that they should find such opportunities. Mathematics is not only science and it is not only experiment and induction. Much of the thrill of mathematics comes from the personal realisation that something *must* be so, and that you, the solver, know why! This is a pleasure that should be enjoyed, as far as possible, by pupils at every level. The fact that they cannot explain everything, and that mysteries remain, is not a fault in the pupil, but an inevitable feature of mathematics which is seldom if every completely clear even to top professionals.

In the equation, $N^2 = 5N + 14$, how fast does N^2 increase compared to $5N$? Why can N not possibly be, say, as large as 20? If the solver knows nothing of negative numbers, why must N be greater than 5, and why can this equation have only one solution, and why *must* every equation of the form $x^2 = ?x + ?$ have exactly one solution? If N is a whole number, why must it be a factor of 14?

3 and 4 are the solutions of $x^2 + 12 = 7x$ because

$$3 \times 3 + 12 = 7 \times 3 \quad \text{and} \quad 4 \times 4 + 12 = 7 \times 4$$

Why do these two patterns naturally go together?

Quadratic equations and real life

Quadratic equations appear frequently in science, at every level. There is one interesting and simple connection with physics, or rather mechanics, which can easily be studied in the mathematics classroom.

A projectile flies through the air in an approximately parabolic arc. This can be seen roughly by observing the shape of the path when a ball is thrown through the air or an arrow fired into the air. However, it is difficult to observe a flight path closely, let alone make a record of it.

A better experiment, and a good simulation of flight through the air, as Galileo realised, is to fix a board at an angle, taping paper to it, and gently flick inked marbles across it. The paths will be uneven but recognisably similar to an inverted quadratic graph. It is an interesting problem to try to fit an equation to the inverted path, or to the path the right way up, once pupils realise that the squared term in the quadratic equation or expression can be negative.

Variations

By starting with a very simple quadratic, pupils have a maximum chance of reaching a solution that is satisfying to them, despite the potential difficulty of the problem. Naturally, however, variations on the original problem will occur to them.

Some variations have been implied already – for example, from considering only integral solutions, or only positive solutions, to allowing fractional or negative solutions.

There are many other possible variations, some very simple, some much more complicated.

What about irrational solutions? If it seems impossible to find such solutions by starting with an equation, how could an equation be deliberately constructed which had irrational solutions? What would it look like? How would it compare with more familiar equations?

Consider multiples of x^2 – for example, $5x^2 - 17x + 29 = 0$. What difference does this make to the numbers of solutions, their relationships, their features? Is it a significant change at all?

Shuffle the terms round in the original version: $N \times N + 7 \times N = 12$. Subtract the 12 instead of adding it: $N \times N - 12 = 7 \times N$. What are the relationships between the solutions of these different equations? How really different are they?

A different variation is to consider equations such as, $N \times N \times N + 12 = 7 \times N$, or $5x^3 - 3x^2 + 41x - 2 = 0$. What properties of the original quadratic are also possessed by such equations? How many solutions can they have? What do their graphs look like? What are the relationships between their solutions?

Computers are ideal for investigating such problems, by allowing rapid study of many different cases. (It is worth remembering that Gauss and Euler were both walking computers before electronic computers were invented.)

Such variations invite the question, 'Are these new equations really the same as the original equation, or are they genuinely different and only superficially similar?' It is an important general concept which applies to all areas of mathematics, that problems which seemed initially different and unrelated, are later perceived to be essentially the same.

Further developments and the use of algebra

'The use of algebra'? This phrase is deliberate. Pupils can, and should, be familiar with a variety of different algebraic notations long before they need to know that algebra is much more than mere notation, however useful. In particular, before they make the amazing discovery that algebraic statements can be manipulated 'almost without thinking' to produce statements, as many as we choose, that are equivalent to the original statement.

All of the problems, questions and variations mentioned so far can be tackled by pupils for whom algebra is at most a code, and many of their results can also be proved by means within their reach.

Fluency in basic algebraic manipulation naturally opens up many more possibilities, especially for proofs and deductions.

If $x^2 + 8 = 6x$, then $x(6 - x) = 8$. The expression $x(6 - x)$ is symmetrical in the sense that if x is replaced by $6 - x$ it does not change. To put that another way, it is the product of two numbers whose sum is 6. That point of view alone is sufficient to allow deduction of all the usual facts about the equation.

Pupils who have spotted the relationship between the coefficients and the sum and product of the roots can prove their conjecture about the sum like this; if the roots of $x^2 - 5x + 10 = 0$ (say) and p and q, then $p^2 - 5p + 10 = 0$ and $q^2 - 5q + 10 = 0$. By subtraction, $p = q$ or $p + q = 5$.

Familiarity with the expansion of $(a + b)^2$ and related expressions allows the calculation of the solutions from their sum and product, when this cannot be done at sight. This is a pedagogically powerful approach because it emphasises the symmetry between the solutions and the coeffients.

If they have come across problems in which an operation such as $x \to 5/x + 2$ is iterated, they will now be able to appreciate that the operation essentially represents a quadratic equation.

Finally, familiarity with products such as $(p+1)(p-8)$ and the squares already referred to, will allow them to appreciate the traditional solutions by factorisation at sight and by completing the square.

Note the qualification 'finally'. Traditionally, pupils have learnt to solve quadratic equations initially *and* finally by completing the square or using a formula. It is also traditional for pupils who can use these powerful devices to have a very shallow understanding of algebra and of quadratic equations and quadratic functions.

In contrast, pupils who have approached quadratic equations from a problem-solving point of view will have an excellent intuitive feeling for what is happening, whether or not they eventually acquire fluency in algebraic manipulation, and will never reach any 'final' point, because they will always by aware of other possibilities, other problems to explore, and questions still to be answered. Which is of course typical of mathematics.

Three additional comments

Richness

Typically, the result of a problem-solving approach is a wealth of opportunities for pupils at many levels, from those who in this present case are fully stretched searching for integral solutions and spotting very simple patterns, to pupils who are soon searching for accurate approximations to non-integral roots, appreciating the value of surds, or manipulating algebraic expressions with confidence.

Typically, also, the variety of questions is matched by the variety of levels of solution. Weak or young pupils who are thinking entirely in terms of integers and who conclude that quadratic equations have sometimes two roots, sometimes 1 and sometimes none at all, are perfectly correct from that point of view. Their conclusion is not a watered-down version of some advanced formula that they do not really understand. It is complete in itself, though it leaves open the possibility that the teacher, or the pupils themselves, will ask, 'Could there be more solutions if fractions were allowed?' or 'Can you find approximate solutions which are nearly correct?'

This is an important point. Such low-level solutions are not incorrect, any more than the traditional O-level solutions are inadequate because they do not encompass solutions through complex numbers or place the solution of quadratic equations in the context of Galois theory.

This richness of possibilities is of immense value to the weaker pupils, but also to the stronger (indeed to all) pupils. I have not emphasised the possibilities open to the strongest pupils because so many teachers, and the 'gifted students' movement, already accept that problem solving is a very valuable activity for such pupils. The possibilities are many, including for example iterative methods, higher-degree equations, and the geometrical properties of the parabola.

My emphasis here has been on the possibilities for the majority of pupils, because it is they who mostly fill our schools, and because they can benefit just as much from a problem-solving approach as the rarely gifted.

True simplicity and false simplicity

Traditionally, going back to the 1950s and earlier, algebra was introduced to pupils by exercises in evaluating $2a$ and $c+d$ and $x-3y+4z$, followed by further exercises in evaluating more complicated expressions, followed by the use of brackets, the solution by standard methods of simple linear equations, and so on, gradually working up to the solution of quadratics by completing the square as a *pièce de resistance* of a secondary algebra course.

This is a perfect example of what I will call 'false simplicity'. Superficially it may seem entirely rational to start with the very simplest exercises and progress steadily through more and more complex material to the final goal.

It is not rational, because it assumes and implies a theory of children's learning which is false, and a theory of what it means to understand mathematics which is also mistaken.

It implies that children are naturally passive receptors of their teacher's knowledge and that their only difficulty is their small cognitive ability to learn from explanation, in deference to which the teacher breaks up the material into very small steps, arranges them in a neat and logical sequence, and explains each step clearly.

It also assumes that understanding of mathematics consists of understanding such sequences of steps. This is also false. Any mathematician's or any teacher's understanding of mathematics is far more complex and subtle than this, involving a multitude of nuances, connections, associations, images, subtleties of meaning, even feelings, which a straightforward explanation will neither evoke nor develop.

Consequently the apparent simplicity of the traditional approach, in every field, from algebra which starts from the simplest little expressions to geometry, starting from axioms so simple that the pupil cannot see the point of them, is an illusion.

Moreover, by denying pupils any personal challenge beyond copying the teacher's explanation, it effectively strips the work of most of its emotional meaning. Whatever cognitive simplicity may be achieved by the teacher's organisation and exposition of the material is more than balanced by the affective vacuity of the material. It totally fails to exploit the pupils' desires to be challenged, to think for themselves, to explore their world.

In contrast, *true* simplicity is found in the richness and variety of problem solving, in which every pupil can find questions which are demanding and difficult, at his or her own level, where the richness has not been stripped away in search of an illusory clarity, and the pupils' experience of solving problems by their own thought processes allows them maximum opportunity to develop their intuitions and feelings for what is happening, which will in turn help to develop a strong and permanent understanding of the problem they have been tackling.

Therefore, in describing how younger and weaker children can tackle the topic of quadratic equations, I am not describing a merely theoretical possibility, but a practical means, not the only one, of introducing pupils to algebra.

The possibility and desirability of introducing such rich topics early on is important for another practical reason. There is no doubt that problem solving, like traditional explanation, can occupy a considerable amount of time. If algebra, to take the present example, is introduced more or less traditionally by a sequence of explanations, each followed by exercises, spread over many lessons, and then a particular topic such as quadratic equations is alone taught through problem solving, then it may be that the total time consumed will be greater than the timetable allows.

By starting with very rich and truly simple situations, the time that would otherwise have been spent on falsely simple explanations is saved, and adequate time for problem solving can be found.

Surreptitious learning

There is another way in which problem solving saves time. Because of the richness and variety of problem solving, pupils are certain to meet ideas which are some way away from the ostensible subject of their problem. In so doing they will make connections and develop ideas which will be of great use to them later. A simple example is concepts of graphical representation which pupils develop from the simplest charts and pictographs and game boards created from primary school onwards. There should be no occasion when pupils meet the idea of a graph or graphical representation for the first time.

The concept of turning numbers into pictures is one of the most general and powerful concepts in all of mathematics, and relates together many different fields, but it is an idea which children should learn surreptitiously long before they come to discuss it explicitly.

Pupils will also, unknown to themselves, be developing more subtle understandings that are more difficult to put into words. There is a traditional tendency to assume that all that teacher needs to convey to the pupils can be put into words, indeed into the words of rather straightforward explanations, and that in doing so the teacher is transferring to the pupil the substance of the teacher's own understanding.

Neither the teacher's nor the pupils' understandings are so simple. Professional mathematicians know full well that it is not enough in order to enter a new area, simply to learn the main results and methods from a book. It is necessary to work in the field, to get a feel for it, to develop intuitive understanding of it, by solving problems and being mathematically active.

Children need just as much to develop their own feelings and intuitions and subtle ideas through their own mathematical activity.

Matrix relating Problem Units to the National Curriculum

Unit	AT1 Number, algebra, measures	AT2 Understand number and notation	AT3 Understand number operations	AT4 Number, estimate and approximate	AT5 Patterns, relationships, sequences	AT6 Algebra: functions, formulae, equations, inequalities	AT7 Algebra: graphical representation	AT8 Estimate and measure, approximate	AT9 Shapes, space and data	AT10 Properties of two- and three-dimensional shapes	AT11 location and transformation in the study of space	AT12 Collect, record and process data	AT13 Represent and interpret data	AT14 Probabilities	Unit
1 Different kinds of numbers	●	○	○		●										1
2 Giant and tiny numbers	●	●	●	●	●	○		●				○	○		2
3 Clock arithmetic	●	●	●		●										3
4 Mental calculation	●	●	●	○	○			○							4
5 Pascal's triangle	●	○	○		●	○									5
6 The Fibonacci sequence	●	○	○		●	○									6
7 The neot number	●	○	○	○	●	○									7
8 Off to infinity	●	○	○	○	●	○				○	○				8
9 Powers and roots	●	●	●	●	●	○									9
10 Rational and irrational numbers	●	●	●	●	●	○									10
11 Sharing and dividing	●	●	●	○	●										11
12 Proportion	●	●	●	○	●										12
13 Measuring in two directions	●	●	●		●	○	○		○		●				13
14 Negative numbers	●	●	●		●	●	●		○		○				14
15 Errors and approximations	●	●	●	●	●	○			○	○	○				15
16 Length, area and volume	●	●	●	●	○			●	●	●	●				16
17 Mensuration, units and scales	●	○	○	●	○			●	●	○	○	○			17
18 Scales and balances	●	○	○	●	○		○	●	●	○					18
19 Percentages	●	●	●	●	●	○									19
20 Diagrams, charts and graphs	●	○	○	●	○	○		●	●	●	○	●	●		20
21 Inventing coordinates	●	●	●		●	○	○		●	●	●				21
22 How fast does it move?	●	○	○	●	●	●	●	●	●		●	●	●		22
23 Making sense of data	●	○	○	●		○	○	●	●		○	●	●		23
24 Averages	●	○	○	○	○	●			●	○	○	●	○		24
25 Combinations and permutations	●	●	○	○	●	○	○		●	○		●	○		25
26 Find the false dice	●	○	●	●	●	○			●			●	●	●	26
27 Combinatorics	●	○	○		●	○			●	●					27

● = Attainment Targets to which the Unit is directly and strongly related.
○ = Attainment Targets to which the Unit is less strongly related.

1 Different kinds of numbers

A secondary headmaster, asked what he wanted to find in pupils coming up from primary school, replied, 'I would be happy if they were friends with the numbers up to 100.'

Easy problems about integers have many advantages. They require minimal previous experience, not even experience of large numbers. (Pupils unused to larger numbers can explore them through such problems.)

They often involve strong visual patterns, and their solution is often aided by the use of the simplest equipment, such as counters or squared paper, as well as calculators, or of course computers.

They are full of surprises and give pupils the opportunity to appreciate the aesthetic dimension of mathematics outside the more obvious and visual realm of geometry. They provide many opportunities for simple but convincing arguments, that is, for elementary proofs. Proofs, for example, that every prime number is one more or less than a multiple of 6, or that the sum of the sequence of powers of 2 is 1 less than the next power of 2, are far easier to appreciate than any proof in elementary geometry.

It is true that many terms are involved, which children need to learn, such as 'odd', 'even', 'integer', 'digit', 'factor', 'common factor', 'multiple', 'product', 'prime', 'composite', 'divisible', 'square number', 'perfect square', 'square root', 'remainder', and so on, but such language can be developed over a period of time, and not one of these terms involves concepts as subtle and difficult to appreciate as, for example, the basic concepts of fractions.

However, the number of terms, and the relationships between them, often means that essentially the same problem can be posed in several different forms. It is true that pupils need to think very carefully indeed about the meaning of many of these problems.

The very variety of the terms used suggests the wealth of simple properties that numbers possess, many of which pupils can explore for themselves. Such exploration helps to develop the friendly familiarity with numbers, including (but going well beyond) familiarity with number bonds and the multiplication table, which is invaluable in later work on more essentially difficult topics.

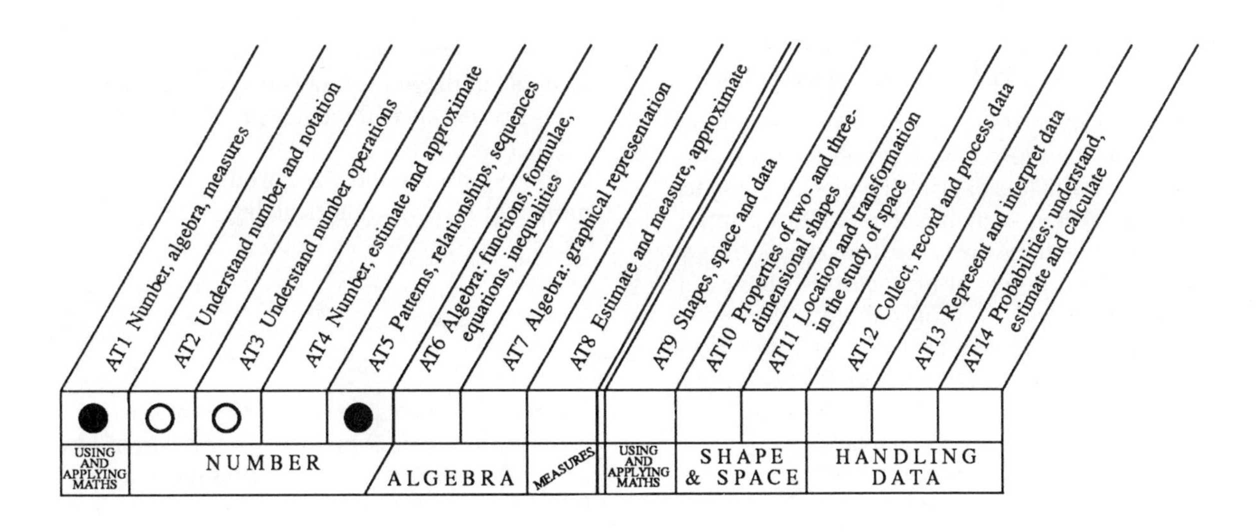

1 Different kinds of numbers

Comments and suggestions

A1 Simply working out the sequence of prime numbers suggests that they become less and less common, but slowly. How slowly?

If there were only a finite number of prime numbers, how many composite numbers could you make by multiplying their powers together?

A different way to look at the number of primes, is to study sequences of consecutive composite numbers. How long can a sequence of only-composite numbers be?

A2 If the problem were posed in numbers written not in base 10, but in binary, or in base 12, then the answer would be different.

Because divisibility by 3 depends on the sum of a number's digits, the digit sum can also be used to test for primality.

All these tests, however, show at most that a number is not prime. Numbers which slip through are not guaranteed to be prime!

A3 The two possibilities are obvious enough from experiment. What is the significance of the number 6 in the problem? What are the possible remainders when a prime is divided by 8, for example, or 12?

How can the conclusion be proved?

This problem is essentially the same as **A2**, because talking about the 'remainder on division by 6' is equivalent to talking about the last digit of the number when expressed in base 6.

A4 How many different prime factors do most numbers have? Because it is easy to construct numbers with lots of prime factors, it is easy to imagine that most numbers have many prime factors. The opposite is the truth. The average number of different prime factors of all the numbers up to 10^{100} is only about 5.4.

This is an excellent example of a problem in which searching for the solution takes a long time, while working out how to construct the solution is very likely much quicker.

How small can a number be which has (say) four different prime factors which are not necessarily different? How small can a number be if it has (say) 10 different factors, not necessarily prime?

A5 What about perfect cubes, or fourth...or nth powers?

What about the prime factors of the product of quotient of two perfect squares?

◆

B1 Is it possible for any number to be expressed as the product of primes in more than one way, using different primes each time?

In other words, is an equation like $7 \times 11 \times 29 \times 101 = 31 \times 23 \times 317$ ever possible?

There is a striking difference between expressing a number as a product and as a sum. 'In how many ways can a number be expressed as the sum of several prime numbers?' is a completely different kind of problem. The special case, 'Can every even number, greater than 2, be expressed as the sum of two prime numbers?' is Goldbach's conjecture, which has never been resolved.

B2 What is the significance of 1 in the question? What about numbers which are 1 more than a perfect square, or 2 less? Or a fixed number more or less than perfect cube or fourth power?

B3 Illustrating products by rectangles of dots or rectangles on squared paper is an early and easy example of the important general concept of transforming an arithmetical problem into geometrical form.

Is it possible to illustrate products of three or more factors by geometrical diagrams?

◆

C1 What kinds of patterns are sought? In the digits of the entries? In the rows or columns? In diagonal lines, of entries a knight's move apart? Between entries which are next to each other, or arranged in a special pattern? As usual, the kind of pattern that is sought, largely determines what is found.

The multiplication table is a good example of adequate richness. In contrast, the addition table has far fewer patterns.

C2 What does 'How many numbers…' mean when experiment suggests that there are an infinite number of them? How many numbers are multiples of 2? How many numbers are square? Or prime?

What is the connection between the three sequences, or sets, of numbers which are multiples of 2, and of 3, and 5?

What if 2, 3, 5 were exchanged for 3, 6, 9?

C3 What is the largest number which the number must be divisible by? What is the relationship between this number and 12 and 15? Pupils who work this out are effectively discovering the idea of an LCM.

C4 What is the smallest number with this property? The largest? The general concept that when a problem has many solutions, there is a pattern between them is well illustrated here. All the solutions form an arithmetic sequence.

Would the problem still be soluble if the remainders on division by 2, 5 and 7, were, for example, 1, 3 and 4 respectively?

C5 The first part may seem obvious when looking at the 7 times table, or the sequence of multiples of 7, but that does not mean that it is unimportant. By looking at an obvious fact in a new way, in terms of 'smallest difference', opportunities for new insights may be found.

For example, it may then be obvious, with no further calculation at all, that exactly one of these numbers is divisible by 7:

244 573 244 574 244 575 244 576
244 577 244 578 244 579

The second question relates directly to **C2** and **C4**.

C6 Problems like this perfectly illustrate the power of very basic ideas. Pupils may well assume that it must be extremely difficult, in view of the large numbers given. Yet it can be solved without factorising any of the three numbers directly, which indeed could have been much larger!

Compare the example at the end of the solution to **C7**.

———— ◆ ————

D1 This beautiful problem was posed by Bruno Brooks on Radio 1. It is a brilliant example of the use of the metaphor of 'colour' to express a problem clearly, which might otherwise be much harder to understand, while making it quite literally vivid in the solver's mind.

The very existence of the words *even* and *odd* makes it possible to talk about members of the class of numbers divisible by 2 and not divisible by 2 in an easy manner. Without such terms, the mathematical abstraction involved would be difficult to hide. There is no equally simple way to talk about typical members of the sets of numbers which leave remainders of 0, 1 and 2 respectively when divided by 3, for example.

D2 What happens in general when even and odd numbers are combined?

What could be said of the cube root, or the nth root, supposing they were integers, of a number divisible by 3?

D3 Can all sufficiently large sums be made up from multiples of 7 and 11? If so, then there will be only a finite number of values which cannot be made up, and also a largest value that cannot be made up.

The behaviour of multiples of 11 in relation to multiples of 3 is relevant here, in the solution to **C5**.

2 Giant and tiny numbers

The Guinness Book of Records is a best seller. Very large and very small numbers are fascinating and mysterious. How many people are there in all the world? How do scientists measure the enormous distances to the stars? How do they know the size of an atom? What do such figures mean?

Even these extreme measurements are trivially small when compared to many of the numbers that occur in pure mathematics. How is it possible to know anything at all about numbers far greater than the number of particles in the entire universe?

Many pupils' intuitive feeling for the sizes of numbers or quantities, large and small, extends only from hundredths to thousands, at the most, while even professional mathematicians have no effective intuition of numbers such as Skewes' number which is $10^{10^{10^{34}}}$.

An intuitive feeling for a much wider range of numbers is not just of practical use, in everyday life and especially in science, but helps to develop an intuitive feeling for the meaning of mathematical concepts and notations, such as that of indices. $10^4 \times 10^5 = 10^9$ is little more than a formal game for a pupil who has no intuitive feeling for numbers that large. (For sufficiently large numbers it must be a formal game, but the intuitive understanding should come first.)

Doing actual experiments and making calculations helps to put giant and tiny numbers in perspective. So do problems in repeated multiplication and division and problems about growth and rates of change. Practical examples such as estimating accurately the number of grains of sand in a jar have the advantage of introducing estimation and error in a very striking manner, as well as inviting pupils to invent their own notations. Consider the popularity of such challenges at fetes and funfairs!

No complicated language is used in this section. In particular, the language of powers and indices has not been used. It can always be introduced if pupils ask how mathematicians represent very large or small numbers, or if that problem naturally arises as a result of pupil's attempts to cope with very large numbers, perhaps by using their own notations.

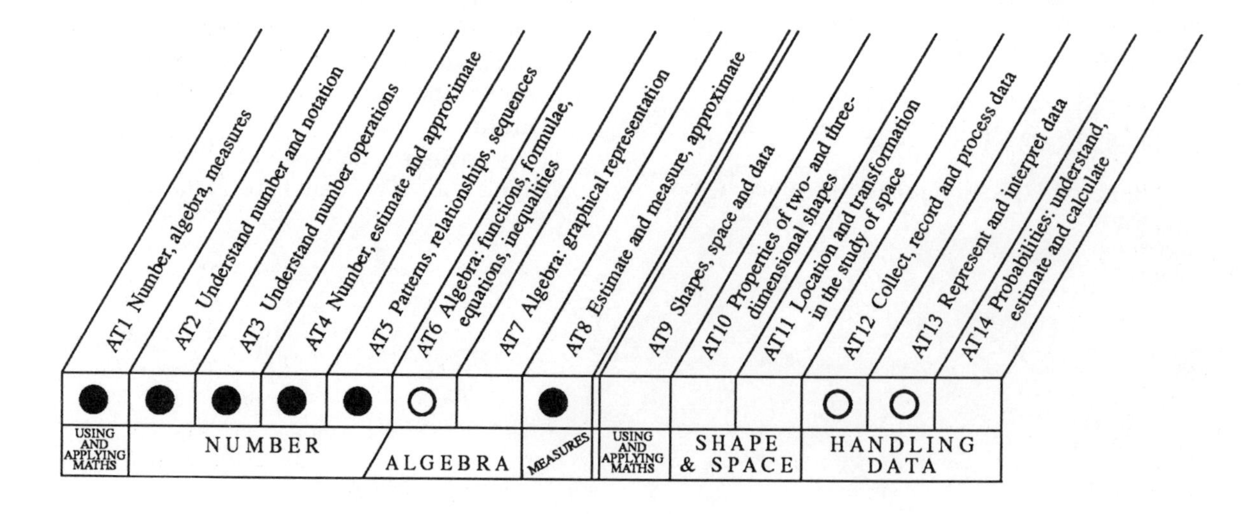

2 Giant and tiny numbers

Comments and suggestions

A1 A wallpapered room may already display a million or more objects, if the pattern is sufficiently detailed and can be separated into individual bits.

A wall sprayed with small dots of paint may contain a very large number of dots, but how can they be counted efficiently? How could pupils know that their number was 1 000 000 and not 1 000 001?

A2 What answers would pupils estimate for these questions, never having made such a calculation before? What would adults typically estimate?

A3 If, as it were, placing two sets of parallel lines at an angle creates an astonishing number of small squares or parallelograms, the effect of dividing up a three-dimensional shape in three directions is even more amazing.

If a metre cube is not available, any cardboard box will do. It is especially surprising how many 1 cm cubes are needed to fill even a small box, because 1 cm cubes are big enough to be easily handled, unlike 1 mm squares.

A4 How many grains of sand on a beach? A classic problem! Archimedes wrote *The Sandreckoner* in which he estimated the number of grains of sand needed to fill the entire universe, which he conceived as contained by the sphere of the fixed stars. He invented his own notation for the giant numbers that appeared. In our notation, his answer was rather less than 10^{51}.

A magnifying glass or microscope is useful for studying individual grains. How much do grains of sand vary in size?

A5 Typical commercial scales may seem to claim, for example from the scales marked, to weigh as little as 1 g, but even if they respond to such a small weight, the response will not be accurate.

Pupils who make their own balances out of light materials and with little friction can detect far smaller weights. Straws and cotton thread are a good start.

A6 Estimate by experiment the time taken to count a single number such as two-hundred-and-thirty-four-thousands-seven-hundred-and-ninety-six. How much time must be allowed for hesitations and errors? What proportion of the numbers less than a million are much shorter to read out than this six-digit number? Are most numbers less than a million six digits long?

It is a significant point that many younger pupils may not be able to count with confidence as high as this.

◆

B1 Problems **A1** and **A3** illustrate the rapid increase in the size of powers of 1000. This problem illustrates the similar increase in size of powers of 2. Why cannot the same effects be achieved by adding numbers again and again?

How does the solution compare with the actual population of England, or of the world, 30 generations ago?

How does the number of great...great grandparents *N* generations ago compare with the total number of great...great grandparents all the way back to that generation?

B2 What about the distances between the sun and the earth?

This example is taken from a popular astronomy book, where such comparisons help to correct the impression gained from physical models of the solar system in museums that the planets are relatively large and close together.

B3 In what sense? The obvious comparisons are in terms of size or weight. Should a comparison of size be in terms of length and height, or volume?

Other comparisons, for example in terms of surface area, are biologically important. The mouse has a relatively larger surface area in comparison to its volume, and therefore loses heat more quickly. If it moved as slowly as an elephant it would die of heat loss. Conversely, an elephant which ran around like a mouse would die of overheating.

There are similar important contrasts in the sizes and strengths of the bones, which depend on their cross-sectional area, which explain why elephants cannot jump!

B4 Compare problem **B1**. A classic answer to this problem is: 'You cannot fold any piece of paper more than seven times!'

How many more times can a piece of A3 paper be folded than a piece of A4 paper? If a piece of A4 paper can be folded *X* times, how big would a piece of paper have to be to increase the number of folds by 10?

The steel for a Japanese sword is folded over itself again and again by the master swordsmith, just as a pastry cook folds pastry again and again, or a potter doubles up a slab of clay. How many layers of steel are there in a sword if the steel is folded 20 times?

3 Clock arithmetic

Clock arithmetic is a fine example of a topic in pure mathematics that is of great importance at a higher level where it is known as *congruence arithmetic* and is important in the theory of number, which generates many patterns, many questions and problems, which are within reach of pupils.

I say 'pure mathematics' because clock arithmetic has almost no practical applications. The 12 or 24 hour clockface is much more a way of understanding clock arithmetic, than the other way round.

Rather, by its contrast with ordinary arithmetic, it prompts in a natural way what would otherwise be very abstract questions about the number system. Problems such as **A5** emphasise structural and abstract features that are not easily questioned within pupils' experience of ordinary arithmetic because they seem too 'natural' to be problematic.

Similarly, the special roles of 0 and 1 are highlighted by problems like **A9**, which also points up a striking contrast with ordinary arithmetic.

The word *modulus* is used in **A8** and **D1**. Pupils find it easy enough to learn provided it is only used as a shorthand for 'the number of hours on the clockface', for example. The rest of the language of congruences is much harder to appreciate because it depends on a new way of looking at familiar ideas, and it is therefore avoided here.

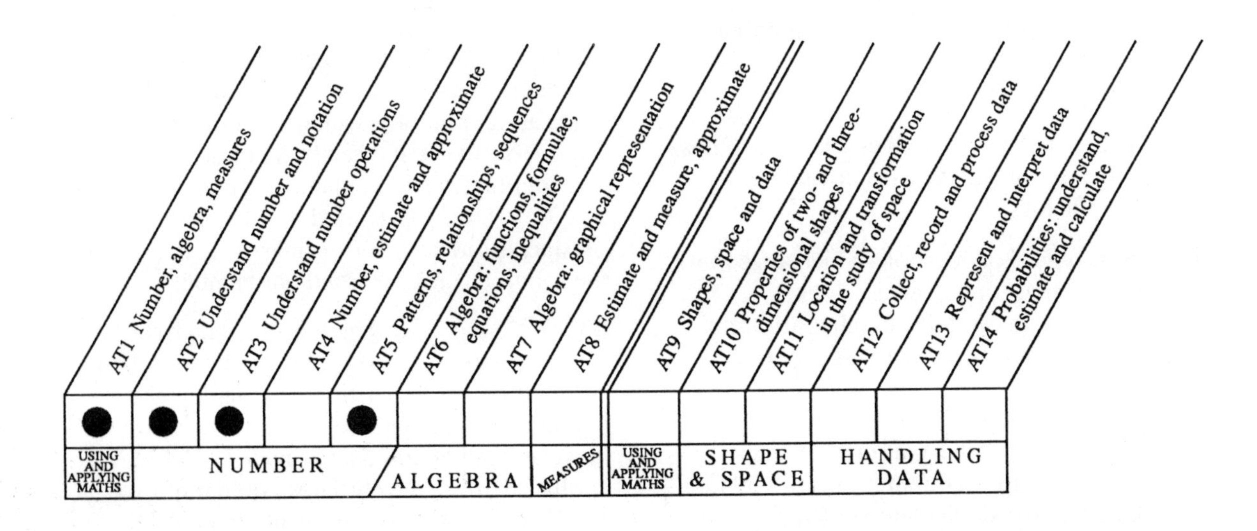

3 Clock arithmetic

Comments and suggestions

A1 Is there any marked difference between the addition tables for different moduli? Is, for example, the addition table for modulus 12 much different for that for modulus 11?

What is the relationship between the addition tables for moduli which are factors/multiples of each other, such as 12 and 24? How many times can the addition table for modulo 4 be found in the table for modulo 8?

A2 How many patterns which can be found in the ordinary multiplication table up to 12 × 12 still exist in the clock arithmetic multiplication table, modulo 12?

Which features of the table depend on special properties of 12, not shared by other numbers?

A3 Is there anything at all about the number 7 which would make its addition table significantly different?

A4 What properties of 12 does 7 lack? What is the significance of 7 being a prime number? Having no factors?

A5 This is typical of questions which arise more naturally in a new topic, such as clock arithmetic, than in everyday familiar arithmetic where the answers are liable to taken for granted.

'Order' is ambiguous here. Is it asking whether you can switch a pair of numbers round, and look up 3 × 4 rather than 4 × 3? Or is it asking if you can calculate 2 × 4 × 3 by calculating either 2 × 4 first, or 4 × 3 first?

Does the order in which numbers are added make any difference in clock arithmetic?

A6 If the answer is yes, do you need to invent negative numbers to go with clock arithmetic? What happens if you count backwards in clock arithmetic?

A7 What difference does it make whether the modulus is prime? When is one third equal to a whole number? Is it necessary to invent fractions to be able to solve all division problems in clock arithmetic, apart from division by 0?

A8 This question can be answered by studying the multiplication table, as can the related questions: how many factors can a number have in clock arithmetic? Can a number be prime in clock arithmetic, having no proper factors?

◆

B1 The idea that the remainder when an expression is divided by, say, 5 will be unchanged if any or all of the numbers in the expression is increased or decreased by a multiple of 5 is typical of ideas which allow surprising conclusions to be drawn about very large numbers which could simply not be tackled directly, and therefore about all numbers of certain types, such as all prime numbers.

Compare the argument of Fermat's theorem, **A4** above.

B2 One of the most important lessons of problems like this is that predictions can be made about numbers which are difficult or impossible to get at directly. So pupils might also consider finding the remainder when (say) 33 525 × 94 377 is divided by 11. The crude product is too large for ordinary calculators, but the solution can be worked out mentally.

B3 $19^{20} + 3$ is a 26-digit number which starts 375...and ends...5604. This can be found on a calculator, using logarithms to calculate the number of digits. To find the entire number, however, and to add 3 and divide by 7 is another matter entirely. How could it be done at all? How long would the calculation take?

It is a remarkable fact that we can predict its divisibility by 7 without knowing any of those facts and of course without calculating the whole number.

Once we have discovered how to 'get at' 19^{20} it is not so surprising that with only a little more work we could discover whether (say) $4059^{100} + 1$ is divisible by 13!

◆

C1 Why must the pattern of remainders repeat eventually? How soon must it repeat?

Why will the sequence of cubes, or fourth powers, or triangular numbers, when reduced by some modulus, go into a repeating pattern?

What conclusions can be drawn from the remainders that appears, or do not appear?

C2 Two-dimensional tables naturally produce two-dimensional patterns which can be represented geometrically so as to please the eye, though this does not necessarily make them easier to understand.

Why will this table inevitably repeat like a tessellation when reduced by some modulus?

4 Mental calculation

It is characteristic of the algorithms (that is, the 'sums') traditionally taught to pupils that they are extremely powerful and wide ranging, and well designed for speedy and efficient use on *large* numbers, or large numbers of small numbers, by Victorian clerks.

They are not designed to display their inner workings, rather the opposite, and they are therefore maximally difficult to understand by pupils.

Moreover, for many calculations with small numbers they are actually inefficient and time-wasting. Pupils who are taught any of the usual subtraction algorithms and then expected to use them to subtract 25 from 43, for example, are being made fools of. The easy way to do this calculation is mentally, by 'bus conductor's' subtraction, counting on from 25. The solution could be '5, to 30, 10 and then 3, makes 18'.

The problems below all illustrate aspects of mental calculation. However, these are not just 'tricks'. Not only are they often useful in practice, but they force pupils to think about what the operations mean, and the relationships between numbers. The fact that $18 = 2 \times 3 \times 3$ is much more significant when this fact is used to multiply or divide by 18 without paper or calculator.

Other properties, such as 'different kinds of numbers' (problem **B2**) can also be exploited for mental work.

The normal algorithms put so much power into pupil's hands (if they understand them) that they actually shortcircuit the pupil's understanding and make thinking unnecessary. Later, when pupils *need* to think (for example in order to understand fractions, ratio and proportion or simple equations) they cannot.

Unfortunately but not surprisingly, many pupils who are accustomed to working everything out on paper have a strong tendency to start mental working by trying to visualise the sum they would normally do, which is usually a difficult if not hopeless approach. They need to know that this is *not* what is expected.

The numbers in these problems have been chosen to be simple. The problem themselves can be thought of as types. For many pupils they could be expressed more generally. Thus **B1** could be expressed: 'Describe an algorithm for adding two numbers from the front.' Other pupils can tackle several variants of one problem in which the only difference lies in the numbers involved. Which choice of numbers make **C3** as easy as possible? When the multiplier is a power of 2, such as 16?

For pupils with a rigid idea of 'sums' and 'correct answers', such variations emphasise that it is the general ideas which are often more important than the specific examples.

All the ingenious tricks of mental calculation depend on important properties of whole numbers. Many other properties suggest related questions. For example, why is the sum or difference of two cubes never a prime number?

As the numbers get larger, pupils may need to jot down partial answers which they would otherwise simply forget. This does not detract from the mathematical relationships they are thinking about. The calculation is still to a large extent mental, and tests the same understanding of basic principles.

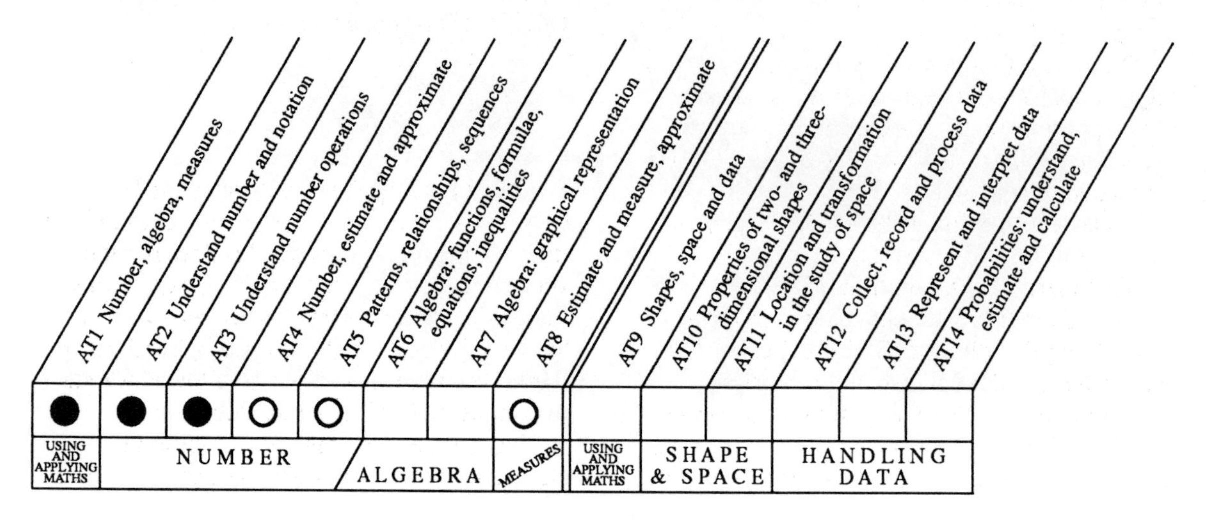

4 Mental calculation

Comments and suggestions

A1 Although addition and subtraction are inverse operations, they are not, as it were, symmetrical, nor are they equally simple. Subtraction, like division in relation to multiplication, is much the more complicated, and it is not surprising that there are many algorithms for subtracting one number from another.

It is typical, and a tragedy, that the usual methods taught in primary schools are neither the simplest to use, nor necessary for many of the sums on which they are practised.

It is fair to say that any child who is expected to subtract 25 from 34 by writing down a sum, like this

$$\begin{array}{r} 34 \\ -\underline{25} \\ 9 \end{array}$$

and decomposing a ten, for example, is being conned. (I appreciate in saying this that a written sum is easier for the teacher to mark than a mental solution — but that is merely to admit that so often the teacher's convenience overrules sound pedagogy.)

The simplest way to solve such a problem is to count on, and this can be done mentally.

Only when subtracting much larger numbers is counting on too much of a strain on the memory, and either requires written working, as in this problem, or should be replaced by the traditional algorithm.

A2 Does this method work for two-digit numbers? Pairs of numbers with different numbers of digits?

There is no practical advantage in this method, but it has its importance in computer algorithms in which it is exceptionally easy to transform a binary number by switching the 1s for 0s and the 0s for 1s.

Its existence again illustrates the variety of subtraction algorithms available.

What methods will pupils come up with if they are given the challenge of inventing their own? A pupil's solution which has been reported more than once in mathematical education journals is to, for example, subtract 329 from 534 place by place, to get the digits 2, 1, − 5, and only then to take 5 from 210 and get the correct answer 205. How natural and simple!

A3 Perhaps the greatest difference between written algorithms and the best mental methods is that the former, division excepted, start from the unit place and work left, and the latter start from the highest place and work to the right.

If this difference is not made clear to pupils, then they may find it natural to attempt every mental calculation from the right, including attempting to visualise a complete traditional subtraction sum, with borrowing and paying back!

If, however, they appreciate that the patterns of written algorithms do not have to be followed, and that starting from the left is allowable, they not only find many mental processes easier, but their answers will make more sense. It is unfortunate that when starting from the right, it is the least significant digits of the solution that appear first, and the most significant arrive last, by which time the pupil may be paying little attention, and a possible absurdity slips through.

Calculations worked from the right are also as hard as possible to remember, because each number is being thought of backwards, with the digits in the opposite order to the sequence in which they are normally read.

Starting from the left, the most significant digits appear immediately. Even if the calculation is not completed, the partial solution will be a good approximation. This is just how pupils should think about calculations.

The particular process of this problem is both simple and useful, although it has no place in traditional arithmetic textbooks.

◆

B1 Adding pairs of numbers from the front is only tricky when two or more pairs of matching digits all produce 'carries'. This does not happen very often. The fact that it can happen means that adding from the front is not, as it were, a 'universal' algorithm. Sometimes it breaks down. (How could the breakdown be allowed for and overcome, with a little extra step?)

This is typical of many non-traditional algorithms. The traditional algorithms work all the time, without exception, but this is no reason for using them all the time. Pupils who naturally add a pair of numbers from the front and only resort to an extra note, or a difficult algorithm on the few occasions when it is necessary, will have more confidence in their own power and flexibility, and better understanding of what they are doing, and will be faster than users of the traditional algorithm.

They will also be much closer in their thinking to the majority of adults, including mathematicians and mathematics teachers, who in practical situations use a mixture of their own methods and tricks, and standard algorithms.

B2 Why is it especially easy to double numbers from the front, rather than multiply them by any other number? This process is very similar to **B1**.

A small but useful advantage of having this facility at one's fingertips, is that powers of 2 are readily available up to 2^{10} and beyond.

◆

C1 Problems like this force pupils to think about the relationship between multiplication and division, and the signifiance of 5 and 2, as factors of 10. (What numbers would be especially significant when counting in a different base?)

Properties of numbers in general are often useful aids in calculation, and, conversely by using such properties in calculation they become more relevant. Compare **D1–3** below.

C2 How many different methods for dividing 282 by 6 are there, including the straight use of the 6 times table and short division?

Division can often be performed mentally with greater ease by spotting suitable 'round numbers'. However, round numbers for this purpose are not usually multiples of powers of 10. What round numbers make this division easier?

C3 There are many more ways to tackle a multiplication problem than a division, and ingenuity is to the fore in choosing the quickest, or the method that is least mentally stressful. Ingenuity, incidentally, should not be despised. Professional mathematicians continually resort to it.

What properties of 18 might be used in the solution? What properties of 132 might be used?

C4 The great differences between multiplication and division are illustrated by the fact that you can multiply by 24 by multiplying by 20 and by 4 and combining the answers, but you cannot divide by 24 in the same way.

Similarly, you could multiply by 25, and then adjust the answer. Is it possible to divide by 24 by dividing by 25 and adjusting the answer?

Pupils can only benefit by being familiar with a number like 24, which is not only 4×6 and 2×12 but also $2 \times 2 \times 2 \times 3$ and $1 \times 2 \times 3 \times 4$ and $5^2 - 1$, among other properties.

◆

D1 Some kind of pattern is not difficult to find. It is much harder to make the description of the relationship as general as possible, and even harder to prove it by an argument, or illustrate it in a picture. The latter idea will only occur to pupils who have had explicit experience of translating arithmetical or algebraic statements into geometrical diagrams.

What happens to the pattern when the numbers are not both even or both odd?

How can the pattern be used for rapid multiplication of pairs of numbers?

D2 What about the relationship between say, 30^2 and 31^2, or between 100^2 and 102^2? How fast do the squares increase, in terms of the differences between them? How fast do differences between entries in a multiplication table increase as you move from the conventional top left corner to the bottom right corner?

Round numbers such as 20 and 30, and small units such as 1 and 2, make this problem much easier to tackle, because the hundreds, tens and units in the product $21 \times 32 = 672$ correspond directly to the products of the tens, tens and units, and units. Typically, there is a large difference in difficulty between a problem using carefully chosen 'simple' numbers, and a logically identical problem using more 'complicated' numbers.

Do the relationships discovered still apply when more complicated numbers are multiplied together?

D3 This problem illustrates the effect of using markedly more difficult language, which makes it much harder for that reason alone. The reader is bluntly told to 'find and *prove*...' a simple rule, and the final qualification is a casual 'and so on for other similar pairs of numbers' which omits to state what 'similar' means in this context.

Pupils' command of mathematical language is exercised and improved by discussion of complex problems such as this. Pupils who have already tackled **D2** successfully will not find this problem difficult, if its meaning is adequately discussed.

Why is it helpful that 9 and 6 have each been increased by one half, rather than some other fraction, or two different small amounts?

5 Pascal's triangle

Pascal himself commented that 'It is extraordinary how fertile in properties this triangle is. Everyone can try his hand' – which is an excellent motto for pupils. There are many simple number patterns to be found in the triangle, which often depend in a straightforward way on the basic rule.

One that isn't included directly in any of the problems below is illustrated by: $56 = 21 + 15 + 10 + 6 + 3 + 1$.

Most of the terms and properties that appeared in Unit 1 suggest questions about Pascal's triangle. Why are there so few prime numbers? Why are all the numbers in the 8th row, except for the end units, divisible by 7? Can a number in Pascal's triangle be a perfect square? (Such questions vary greatly in difficulty!)

Pascal's is not the only notable triangle. Leibnitz' triangle has interesting properties:

Each fraction is the difference between the two fractions above it and next to it. However, it is notable, and typical, that the original is richer and more important than any of its variants.

Pascal's triangle, and Leibnitz's, offer pupils opportunities to invent their own notations for particular numbers. Without a sensible notation, how can they be referred to conveniently?

There are much more advanced uses of the triangle in combinations and permutations and in expanding $(a + b)^n$, for example, which rapidly become much more difficult. These are only hinted at in **C2** and **C3**. Some problems appear in Unit 25 (pp. 171–7).

$$\frac{1}{1}$$

$$\frac{1}{2} \quad \frac{1}{2}$$

$$\frac{1}{3} \quad \frac{1}{6} \quad \frac{1}{3}$$

$$\frac{1}{4} \quad \frac{1}{12} \quad \frac{1}{12} \quad \frac{1}{4}$$

$$\frac{1}{5} \quad \frac{1}{20} \quad \frac{1}{30} \quad \frac{1}{20} \quad \frac{1}{5}$$

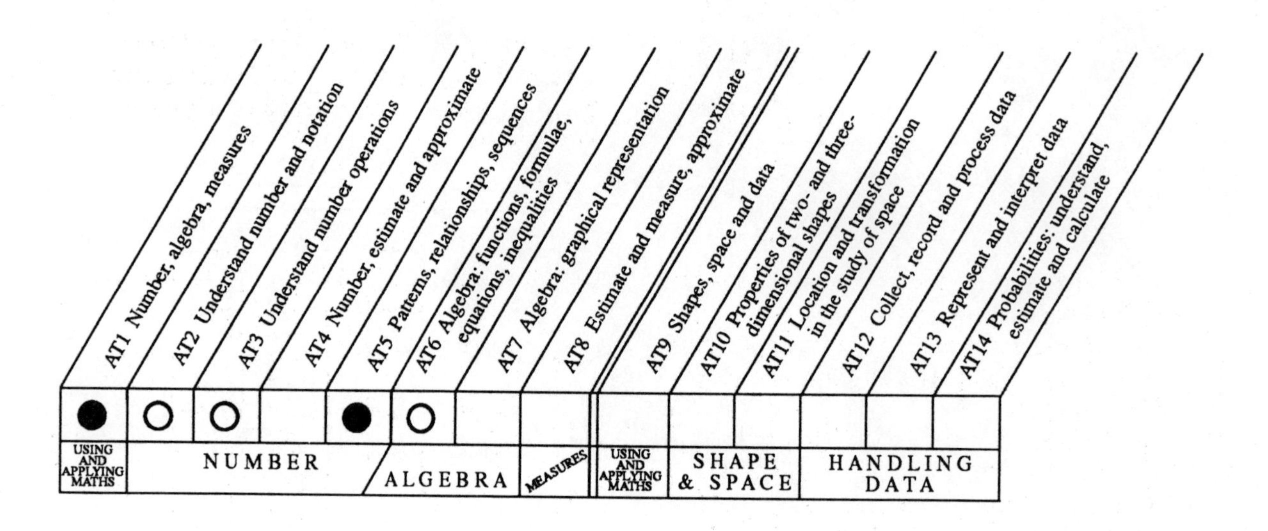

5 Pascal's triangle

Comments and suggestions

A1 The pattern is sometimes presented with the topmost 1 missing. If it is present then it makes mathematical sense to count it as the zero row, and the 1-1 row as the first row, which is another illustration of the curious role of 0.

The pattern suggests many problems, some of them extremely difficult. Most numbers appear twice in the pattern. Can a number, apart from 1, appear more than twice? Which is the first number, apart from 1, to appear several times? What is the pattern in the central column taken in isolation? And so on.

A2 The sums of the rows make a strong pattern. How can this be predicted from the 1s at the ends of rows and the basic rule for the entries that are not ends of rows?

A3 Any array of numbers suggests problems about the sums of rows, or columns, or the entries within a region, which are analogous to the sums of ordinary sequences. Since Pascal's triangle is defined row by row, it is natural to ask for the sums of rows, and to expect a relatively simple answer.

How will this answer relate to the previous problem?

A4 The much earlier Chinese version of the triangle was arranged like this:

```
1
1  1
1  2  1
1  3  3  1
1  4  6  4  1
..........................
```

In this arrangement the normal diagonals sum to the Fibonacci sequence 1, 1, 2, 3, 5, 8,.... Pascal's arrangment makes these diagonals into knight's moves diagonals, as it were, and very difficult to see.

All sets of entries along a straight line in Pascal's triangle will inevitably show a pattern. This pattern and the patterns of rows which are parallel to the outer rows of 1s are merely the simplest. See problems **D1–4**.

B1 The pattern is much stronger when it is drawn for a large number of rows. (A computer would be ideal for that task.)

Colouring numbers as odd or even is effectively showing their remainders on division by 2. Problems **B1–3** are, as it were, problems about Pascal's triangle in clock arithmetic. As usual, a pattern appears when any sequence or array is reduced to some modulus.

Where do the inverted triangles in the pattern come from? How can they be predicted from Pascal's basic pattern?

B2 The same questions may be asked of this pattern: why does row 7 contain only multiples of 7, the ends excepted? where does the inverted triangle shape come from?

Is it relevant that 7 is a prime number? (The numbers of the rows in the last problem 1, 2, 4, 8, 16,...are all powers of the prime number 2.)

B3 The number 3 is too small to show a strong pattern in the first few rows. The pattern is stronger when the $3 \times 3 = 9$th row appears.

What is the connection between the patterns for 2, 3 and 6?

What proportion of Pascal's entries are multiples of 3? Are there as many which are 1 less, or 1 more, than a multiple of 3?

Are even entries as frequent as odd entries?

C1 How many ways are there to get from the top left dot, to the marked dot in this square array, which could represent a very regular street plan?

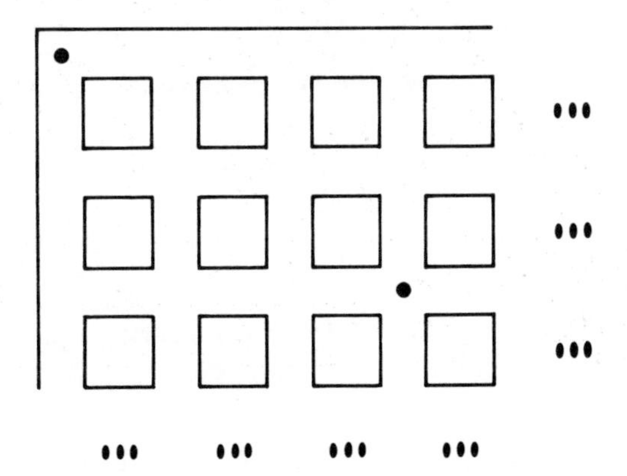

Pascal's triangle can be drawn like this – a third way of looking at it! How does it relate to the grid of dots and lines?

1	1	1	1	1	1	...
1	2	3	4	5	6	...
1	3	6	10	15	21	...
1	4	10	20	35	56	...

....

C2 This problem is a simple example of a problem in combinations, or selection, or choice. The solution to any problem of the kind, 'how many ways are there of choosing X different objects from Y different objects?' appears in Pascal's triangle, provided the order in which the objects are chosen does not matter.

C3 All such products in algebra can be thought of as the result of selecting terms from the backeted expressions. They are therefore related to problems of selection, and to Pascal's triangle.

It is not so difficult to calculate the product, for those pupils familiar with products of two expressions. It is much harder to predict the result by a general argument.

In how many ways can the term t^2 appear in the final product?

◆

D1 Pupils will very likely have come across the triangular numbers already, in which case they are there to be recognised along an obvious diagonal.

Since the triangular numbers in one diagonal can be represented so simply in a pattern of dots, how might the numbers in the next diagonal, 1, 4, 10, 20,... be represented?

D2 Piling table-tennis balls into regular tetrahedra produces the sequence in the next diagonal, 1, 4, 10, 20,....

This is, notably, a three-dimensional pattern, in contrast to the triangular numbers which make a two-dimensional pattern. Moreover, it is not difficult to see why it must be, in a sense, three-dimensional, because each layer of a triangular pyramid of table-tennis balls is triangular, so each tetrahedron contains the sum of the first so many triangular numbers....

What kind of pattern would be expected from the sequence 1, 5, 15, 35, 70,...?

D3 The triangular numbers have a wealth of properties. Some are analogous to the properties of the tetrahedral numbers and other sequences of numbers which appear in Pascal's triangle. Others match the properties of the square, pentagonal, hexagonal numbers.

Triangular numbers are especially suitable for pupils to explore because they can be represented by triangular patterns of dots, which do not necessarily have to be in the shape of equilateral triangles, and which can be shuffled around, dissected or assembled, much like geometrical triangles.

They also invite pupils to invent their own notations, so that they can refer to the 'third triangular number', for example, as briefly as possible.

In the present problem: how can a square be made up from triangles?

D4 Some pupils are as likely to be as attracted by the idea of four dimensions as others will be put off. Higher dimensions are like giant and exotic numbers, exciting provided they do not cause panic.

Pupils who follow the analogy from triangular numbers, represented by plane patterns, to tetrahedral numbers represented by three-dimensional arrangements, may well suspect that the next diagonal, which starts 1, 5, 15, 35,... can be thought of as four-dimensional. Not easy to draw in a picture – but what about the kind of formula for the numbers in a 'four-dimensional' sequence?

6 The Fibonacci sequence

Apart from its historical interest, Fibonacci's sequence is extraordinary for its wealth of simple and elegant properties, only a few of which appear in the problems.

It illustrates very well some general ideas that are not so apparent from most other sequences, even sequences as apparently simple as the squares. The idea for example that there is a formula for the sum of the terms, or the sum of the squares of the terms, as well as relationships between adjacent terms: that an initial simple pattern should give rise to many other patterns is obvious to the mathematician, but has to be learnt by children.

The same kinds of questions can be posed about the sequences in the next two entries, but the patterns and results are usually far more complicated. There are several senses in which the Fibonacci sequence is exceptionally simple.

Like Pascal's triangle, it almost requires that pupils invent a suitable notation so that they can refer to the fifth term, or the eighth term…more briefly. Whatever they decide upon, it will be an example of one important aspect of algebra, strikingly different from the use of letters to stand for unknown numbers or as general variables.

If pupils are not familiar with any algebraic notation, sequences like this give them a chance to use their own notation in an algebraic way which makes perfect sense in context.

If pupils are familiar with elementary algebraic manipulation then they can investigate much more complicated properties, and prove their results.

Fibonacci's sequence also has a curious and unexplained relationship to the botanical phenomenon of phylotaxis. The leaves of a plant grow in a spiral round the stem, so that the angles between successive leaves are constant. The commonest angles are 180°, 120°, 144°, 135°, 138°27′, 137°8′,…, which when expressed as ratios of the whole circle are $\frac{1}{2}, \frac{1}{3}, \frac{2}{5}, \frac{3}{8}, \frac{5}{13}, \frac{8}{21}, \ldots$, the ratios of alternate members of the Fibonacci sequence.

The sequence also occurs in the surface pattern of pine cones and pineapples, for example, and in the pattern of the florets of a sunflower.

On the other hand, computer scientists use Fibonacci's sequence to construct efficient algorithms for searching tables of data. Contrary to appearances, the sequence is of much more than merely historical or recreational interest.

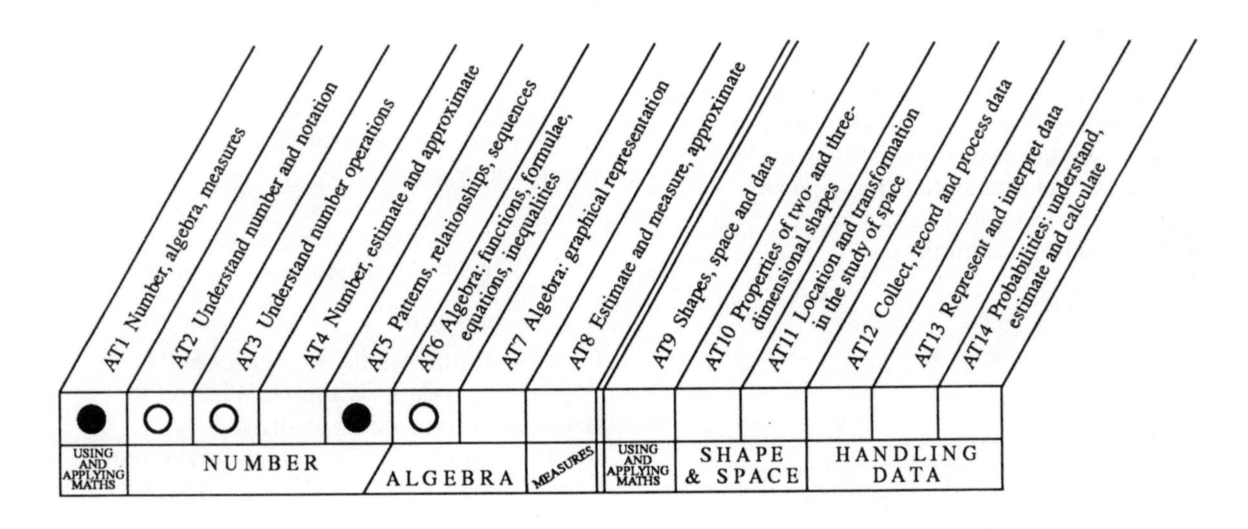

6 The Fibonacci sequence

Comments and suggestions

A1 Fibonacci himself started his sequence 1, 2, 3, 5, 8,..., supposing that the first pair bred immediately.

Drawing a neat diagram of Fibonacci's rabbits for the whole 12 months is not easy. It is also unnecessary, because there is a simple pattern which allows the number of rabbits at the end of each month to be written down very quickly.

This is a typical example of a practical situation in which the mathematician rapidly gets rid of most of the initial conditions, only keeping the absolutely essential.

A2 It is a historical coincidence that Fibonacci's problem leads to such a very simple rule.

What other rules of breeding might be used? Alternatively, focussing only on the rule of Fibonacci's sequence, what other similar rules might be used to generate a sequence?

A3 Children have a natural tendency to look at the differences between the terms of a sequence, which is fortunate, because this is one of the most powerful means of analysing a sequence, chiefly because if there is a polynomial formula for the terms of the sequence, then the formula for the terms of the sequence of differences will be one degree lower.

Thus, the differences in the sequence of perfect squares (degree 2) is the sequence of odd numbers (degree 1). The fact that the sequence of differences of the Fibonacci sequence is not simpler than the original sequence, proves that it cannot have a polynomial formula for the terms.

B1 Summing the terms of a sequence is another natural tendency, though the result is usually more complicated than the original sequence. (It is essentially the opposite process to finding the differences.) For example, the formula for the sum of the first n perfect squares (degree 2) is a polynomial of degree 3, and the formula for the sum of the first n cubes is of degree 4.

However, we might expect the Fibonacci sequence to be a little simpler, because after all the rule for creating it is to add two of the terms together. It would not be surprising to find a strong and simple pattern by adding all the terms.

B2 This property might be spotted by mentally multiplying each term by the next term but one, and spotting a connection with the original sequence.

What other relationships are there between the products of terms in the sequence?

B3 It is not surprising that there are relationships between squares of the terms and other products of two terms, or between the sums of the squares of the terms. However, this is not because squares are especially related to Fibonacci; there are relationships between the cubes of the terms also, and higher powers. Rather, squares are the simplest powers we can consider, apart from the terms themselves, and naturally patterns involving squares are much easier to spot.

B4 This problem gives the solver a large hint as to how the formula for the sum of the squares of the terms might be found. It makes up for this generosity by demanding a 'proof'. The most important feature of such a question for most pupils is not the geometrical diagram as such, but their understanding of what a proof entails.

What do they need to say, and how, in order to make quite clear to someone reading their proof, that the result they are claiming to prove is crystal clear and absolutely certain, beyond any doubt?

C1 Even the divisibility properties of the sequence are relatively simple. What is the sequence of remainders on division by 2? Why must the sequence of remainders on division by 5 repeat, sooner or later? How soon does it repeat?

C2 The remainders on division by 7 will show a pattern, which will reveal, among other facts, which terms in the sequence are multiples of 7.

Is it significant that 5 and 7 are primes? What patterns will appear when the terms are reduced to a modulus that is not prime?

C3 Is it possible to predict which terms, or at least one term, which will be divisible by a given prime p?

The result of **C2** is a good hint here, though the result of **C1** is not: the number 5 has a special and idiosyncratic connection with the Fibonacci sequence.

C4 This is another opportunity for pupils to appreciate the idea of proof, and the power of proof. The conclusion follows from the basic Fibonacci rule, plus the fact that if two numbers have a common factor, then their sum or difference has the same factor. Yet it says a great deal about all Fibonacci numbers, however large.

———◆———

D1 Patterns such as $8 \times 21 = 13^2 - 1$ state that $8/13 = 13/21$, approximately.

In other words, the ratio of pairs of successive terms is approximately the same, and so the sequence is roughly geometrical. Which geometrical sequence fits it most closely?

This is a very different situation from the square or triangular numbers. In these sequences, the ratios of successive terms do become roughly the same — but only as the ratios tend to 1. The ratio of any pair of sufficiently large consecutive squares is close to 1, and the same is true of consecutive triangular numbers.

D2 The ratios of successive terms, by calculator or computer, are approximately:

1 2 1.5 1.6666... 1.6

1.625 1.61538... 1.619047...

This sequence illustrates the power of direct calculation and its weakness. This sequence seems to be clearly tending to a limit, and consecutive terms are respectively greater than and less than the limit, whatever it is. It is very typical that a limit should be approached from above and below alternately.

But what is the limit? How can it be expressed more precisely?

An additional query: what about generalised Fibonacci sequences, using the same rule, but starting with a different pair of numbers – do the ratios of their terms tend to a limit?

7 The next number

Sequences have a special fascination for pupils, on account of their strong patterns, and their mysterious destination, infinity.

They provide many opportunities not only for spotting patterns and connections, but also for simple arguments and proofs.

The term 'rule' is used very sparingly in this book, in relation to sequences. It is not inappropriate when pupils make up their own rules for a sequence, though even then the language of patterns or relationships or formulae is at least as satisfactory. It is misleading when pupils are studying a sequence to find 'the rule' by which it is constructed, because there never is only one such 'rule'. Rather, there are different ways of looking at it, and all these different ways may not be consistent with each other. This is especially so when the sequence is short. Different rules may lead to quite different continuations.

The numbers in these sequences are very simple. Some pupils may prefer to handle harder numbers, perhaps with the use of a calculator. This can be a means of understanding harder numbers. For example, instead of waiting until pupils are familiar with decimals before studying sequences involving decimals, studying the latter can be a way of giving pupils experience of decimals in a meaningful context.

All sequences can be represented on a graph. However, only some properties of a sequence will be illuminated by a graph; many properties may be hidden complete.

A surprising number of sequences can be represented very neatly by geometrical diagrams. Such diagrams often provide opportunities for simple arguments and proofs, as well as developing the general concept of a strong link between geometry and arithmetic or algebra.

Algebra is not mentioned in this section, but it is easy to see how algebra can be used as a notation when talking about sequences and how algebraic manipulation can be used for discovering and proving properties of sequences.

Lack of experience of algebra does not mean that pupils cannot use what is essentially algebraic notation. When pupils are encouraged to invent their own notations they are already using algebra as a language. (It is worth noting that the traditional algebra of generalised arithmetic does not include at school level the use of subscripts, although these are naturally useful in referring to terms of a sequence.)

If pupils end up talking about S10 + S11 or 10* + 11*, rather than the '10th term added to the 11th term', then they are already using algebraic shorthand in a mathematically realistic manner, which will aid their future understanding of any kind of algebra.

Once pupils have studied some sequences and found some of the patterns for themselves, making up their own sequences and giving them to each other to solve, or to the teacher, is an excellent activity, though pupils are liable to invent the most obscure rules they can think of.

Interesting and important arguments arise when a sequence is perceived to have more than one possible pattern. Are the patterns equivalent? Can this be proved? Or are they genuinely different ways of looking at the same sequence, leading to different continuations of the sequence? Such arguments are extremely instructive.

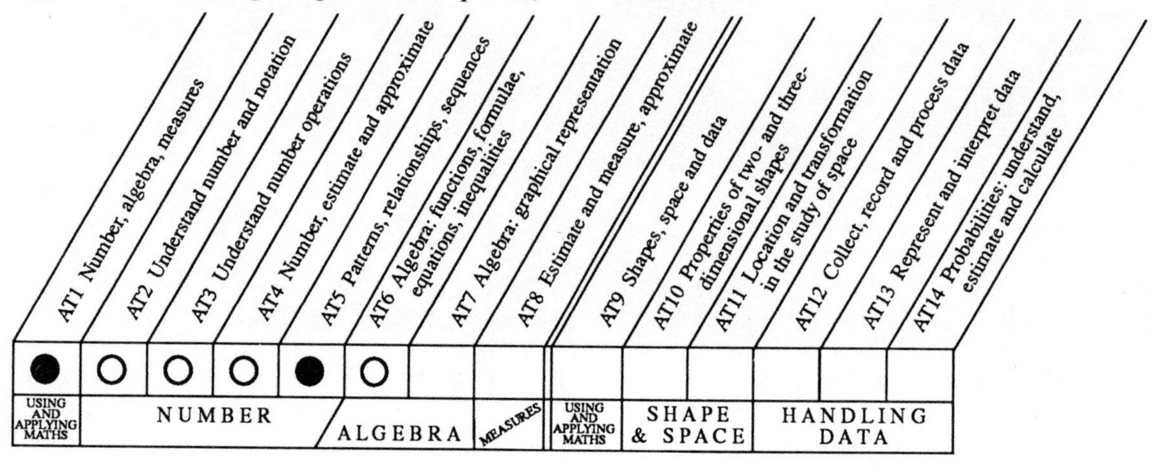

7 The next number

Comments and suggestions

A These problems are just five of the vast number of plausible sequences that might be considered.

The question asks 'How do you think…' deliberately to emphasise that there is an element of pupil choice, though even a professional mathematician might need to think for a moment or use paper and pencil to come up with several different patterns for **A1**, in which six terms are already given. **A5** is another matter. How many simple patterns could fit this sequence of only three terms?

A1 Pupils naturally tend to look at the differences between successive terms. If the sequence of differences does not seem very simple, what about the sequence of differences between the differences?

Just looking at the size of the terms may give a hint to pupils with a little experience of sequences. How fast are these terms increasing?

A2 This sequence may seem to involve two different patterns, one for the 1st, 3rd, 5th,… terms and the other for the 2nd, 4th, 6th…terms. This is typical of many patterns that pupils make up for themselves, which they delight in making complex.

A3 The pattern of differences is very strong. A perfect example of a sequence in which it is helpful to look at how fast the terms are increasing. How does the rate of increase of the terms of this sequence differ from the rate of increase in **A1**, which is another rapidly increasing sequence of terms?

A4 The pattern of differences is very simple. However, this is not sufficient to tell you much about the sequence, apart from its continuation. What other properties does it have? What are the connections between these properties and the pattern of differences?

A5 A sequence with very few terms is like a coded message of only a few words. In a sense, it is easier to 'solve' such a sequence, because there are more possibilities. On the other hand, you are given less information, so you have less to go on, and you need to use your own imagination more.

How many different plausible sequences start 1, 2, 6? How many different rules for such a sequence are equivalent to each other, and lead to the same continuation, and how many lead to different continuations?

———◆———

B1 Number patterns such as **B1** and **B2**, by showing how the numbers in a sequence are built up, are a very concrete way of talking about sequences, which emphasises the importance of the differences between the terms. They also emphasise the general concept that one pattern leads to another, that you can expect the partial sums of a sequence to show a pattern themselves.

How does the right-hand sequence compare with the sequence of **A3**? What patterns will there be in the sums of powers of 5? Do all sequences of sums of powers have essentially the same pattern? How can the pattern be proved from the pattern of sums on the left?

B2 In contrast to the sequence of **B1**, this sequence, and its partial sums, increases relatively slowly. How fast do the partial sums increase?

How does this sequence compare with sequences such as

$2 + 6 + 10 + 14 + 18 + \ldots$?

How can the symmetry in the sequence $1 + 4 + 7 + 10 + 13$ be used to find its sum?

———◆———

C1 Sequences typically have simple divisibility properties, though not usually as simple as those of the Fibonacci sequence in the last unit. This problem introduces the idea that something does not exist, or that something is impossible. In this case, it is impossible to find a term in this sequence, however far you go, which is a multiple of 3.

Proofs of impossibility are often attractive to pupils, because of their surprise value and their power.

How many numbers in an arithmetical sequence of integers whose common difference is 5 would be multiples of 3?

C2 Another example of impossibility. What is the natural pattern for this sequence? What property of perfect squares is needed to show that there are none in the sequence?

C3 These are the triangular numbers, whose formula is $\frac{1}{2}n(n+1)$. What will be the pattern of divisibility by 5 for other quadratic functions?

C4 'Being the sum of two perfect squares' is a complicated property, much more complicated than divisibility properties. However, it can easily be investigated by pupils. Which numbers are sums of two squares? Which are not? Given two numbers which are, what other numbers can you predict will be the sum of the two squares?

◆

D1 Could the sequences be equally well illustrated if they started with different numbers, that is, for example, if each term was increased by 1, and the sequences each started with a 2?

Can the sum of all arithmetical sequences be found by using suitable geometrical diagrams?

How can the sum of a power series such as $1+2+4+8+16=31$ be illustrated by a diagram? Compare the diagram for the sum of the squares of the Fibonacci numbers.

8 Off to infinity

The idea of infinity is both curious and intriguing and difficult to understand, and yet somehow familiar. We talk in everyday language of 'going to the limit', and the idea of infinite, or at least unbounded, space is more familiar than it used to be. Yet many pupils may be surprised by the idea that it is possible to add up the terms of a sequence 'for ever and ever'.

This is hardly surprising. Professional mathematicians were for a long time confused as to the meaning of such an 'infinite sum', and the idea of infinity continues to provide mathematicians with many of their trickiest problems, as well as many of their most elegant and beautiful ideas.

Problems about infinite series can provide pupils, also, with examples of elegant and attractive mathematics.

Unlike the integer sequences in the last unit, some facility with fractions is necessary to handle infinite series converging to a sum. This facility, however, need not be great. An elementary understanding of fractions is sufficient, and, of course, the experience of using their understanding of fractions to tackle tricky problems will itself develop their understanding of, and intuitive feeling for, fractions.

As in Unit 7, opportunities abound for pupils to use their own notations, and to use them in an essentially algebraic way, developing feeling for algebraic notation on the way.

The sums of many series can be illustrated by geometrical diagrams which show vividly how the sizes of the terms decrease, and develop intuitive understanding of how it is possible, for example, for an infinite number of terms to sum to a limit, and what the idea of a limit means.

Sequences and series show up the good points and the bad points of calculators and computers. Finding an approximate sum, very quickly as an experiment is often a very suggestive first step, but without any pattern to study the sum may be unrecognisable (compare the appearance of $\frac{1}{2}(\sqrt{5}+1)$ in Unit 6, problem **D2**). Even if recognised, no reason for the limit may be apparent from the mere figure.

In contrast, working out partial sums, expecting to find a pattern in them, produces excellent material for pattern spotting and simple chains of reasoning, simple proofs, while for certain series other kinds of logical arguments arc available to demonstrate the power of argument over experiment alone.

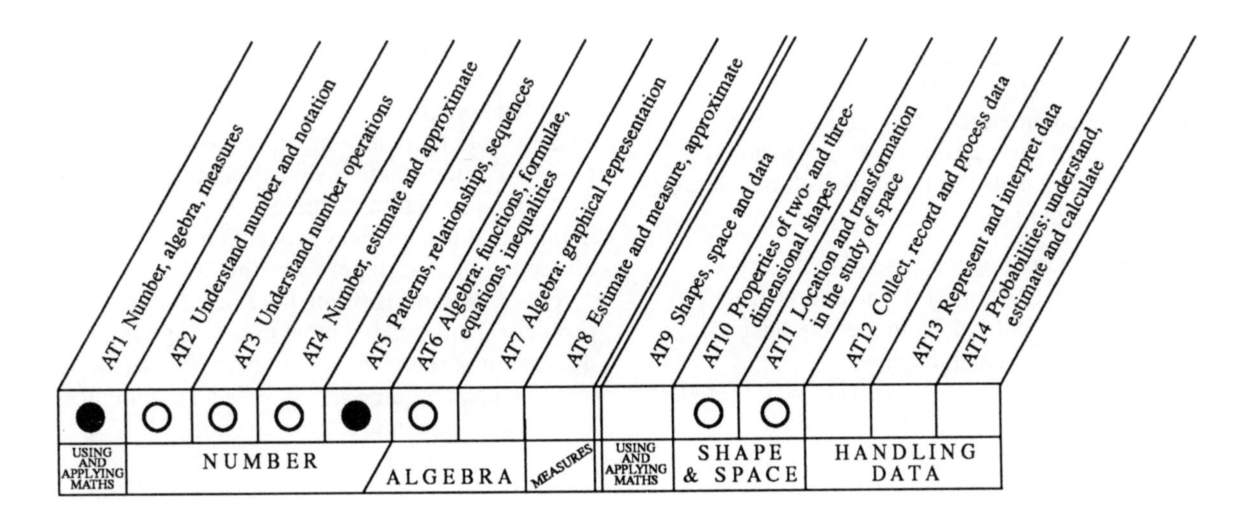

8 Off to infinity

Comments and suggestions

A1 The simplest series generally correspond to the simplest integer sequences. The powers of any number, starting with 1, make a simple sequence or series, and as usual the powers of 2, or of $\frac{1}{2}$, are the simplest of all.

In particular the sequence of partial sums is as simple as could be.

How can the partial sums, and the sum of the whole series be illustrated in a diagram?

The problem asks for the sum that the solver 'expects', because it does not assume that the idea of the sum of an infinite series is familiar. Examples such as this make the idea plausible, because they are so intuitively plausible themselves.

A2 What are the patterns of partial sums of these series? How do these patterns generalise to the series of reciprocals of powers of the numbers?

The series of powers of $\frac{1}{4}$ starts with a small fraction and the terms rapidly get much smaller still. It is relatively easy to believe that such a series has a limit.

What about the series
$1 + 0.99 + 0.99^2 + 0.99^3 + \ldots$
whose terms get smaller very slowly?

A. There are many possibilities, both one- and two-dimensional. The clearest solutions are as much a matter of good artistic design, as anything else. How can the undoubted mathematical facts be presented as clearly as possible? For this reason, a two-dimensional diagram is often clearer than a one-dimensional figure, though in a sense the latter is 'simpler'.

A4 This is a kind of converse of summing a series. However, actually trying to construct $\frac{1}{3}$ by choosing or discarding from the sequence $\frac{1}{2}, \frac{1}{4}, \frac{1}{8}, \frac{1}{6}, \ldots$ is quite different in feeling.

Can you be certain that $\frac{1}{3}$ can be constructed at all? Might there not be a stage at which adding the next term would take the total over $\frac{1}{3}$, but omitting it would leave a series of terms remaining whose total sum was not sufficient to reach $\frac{1}{3}$?

In how many different ways can $\frac{1}{3}$ be constructed from these terms?

How does this problem relate to the problem of expressing $\frac{1}{3}$ as a binary decimal?

A5 The simple integer sequence 2, 6, 12, 20, 30, ... appeared in Unit 7. Since all the terms are even, this series is one half of the sum of the reciprocals of the triangular numbers:

$$\frac{1}{1} + \frac{1}{3} + \frac{1}{6} + \frac{1}{10} + \frac{1}{15} + \ldots$$

What is the sequence of partial sums?

A6 The terms in this series decrease quite quickly to start with (from $\frac{1}{2}$ to $\frac{1}{3}$ is a drop of one third,) but their rate of decrease gets slower and slower the further the series continues, in contrast to the sequence, for example, of powers of 0.99, in which every term is 0.99 times the previous term.

This is the kind of series, getting smaller but more slowly, which might possibly not have a limiting sum at all. How could a target total, of (say) 10, be reached and exceeded?

What other sequences have a similar property of decreasing term by term, but more and more slowly?

———◆———

B1 The form and language of this problem is sufficient to indicate that it is difficult, or rather that the poser thinks it is difficult for the intended audience. The form of the argument suggests that there is some motivating idea behind it, but nothing is said about this idea. The bald instruction 'complete this argument' implies that the solver is expected to follow and understand the argument without help.

This seems appropriate here because the problem *is* difficult for most pupils, for whom a discussion of the form of the very idea of a mathematical argument, and how an argument is laid out and how much you can expect the readers to work out for themselves will be more valuable than an attempt to solve the problem as stated.

How much do pupils assume when they write out an explanation of one of their own arguments? Do they think about the person they are writing for? Or are they writing for themselves, and perhaps their teacher?

B2 Formal manipulation leads to absurd results when false assumptions are innocently made along the way. In this case, is it reasonable to call the sum of the series *S*, or anything else, when the sum does not exist?

B3 The question 'how many?' when asked of the members of an infinite set can be ambiguous. This is a typical example of a mathematical phrase which is unambiguous at one level, suddenly becoming tricky and ambiguous at a higher level, where is has to be sorted out by some clear thinking.

What do pupils think it ought to mean in this case?

Do all infinite sequences contain the same number of terms?

What should it mean when comparing these two sequences?

1	2	3	4	5	6	7	8	...
2	4	6	8	10	12	14	16	...

Does the first sequence contain twice as many numbers as the second? In what sense, since they both contain an infinity of numbers?

◆

C1 The idea of a limit does not apply only to sequences of numbers. What is the limit of a sequence of rectangles whose area is constant, but whose breadth is decreasing?

C2 In some respects two-dimensional sequences like this geometrical spiral are easier to understand than an ordinary sequence, because it is easier to follow where they are going with the eye. (Here, consecutive arms are perpendicular and their lengths are the sequence $1, \frac{2}{3}, \frac{4}{9}, \frac{8}{27}, \ldots$)

Pupils with access to LOGO can easily explore such spirals, and other sequences and limiting processes, though as usual experiment by itself is not enough to give all the answers.

They are easy to design, because the rules for the change in length of the vector and for the change in direction can be entirely independent of each other. In this problem, the angles between the arms could be changed (they do not have to be equal anyway), and the lengths of the arms could be decided according to any rule the pupil chooses.

What kinds of rules lead to a sequence of vectors with a limiting point? What kinds of rules lead to sequences of vectors that repeat? Or go off to infinity without tending to a limit or repeating?

C3 The angles of this spiral are 60°, the next simplest after the right angles of the previous problem.

◆

D1 What is the rule by which this sequence is constructed?

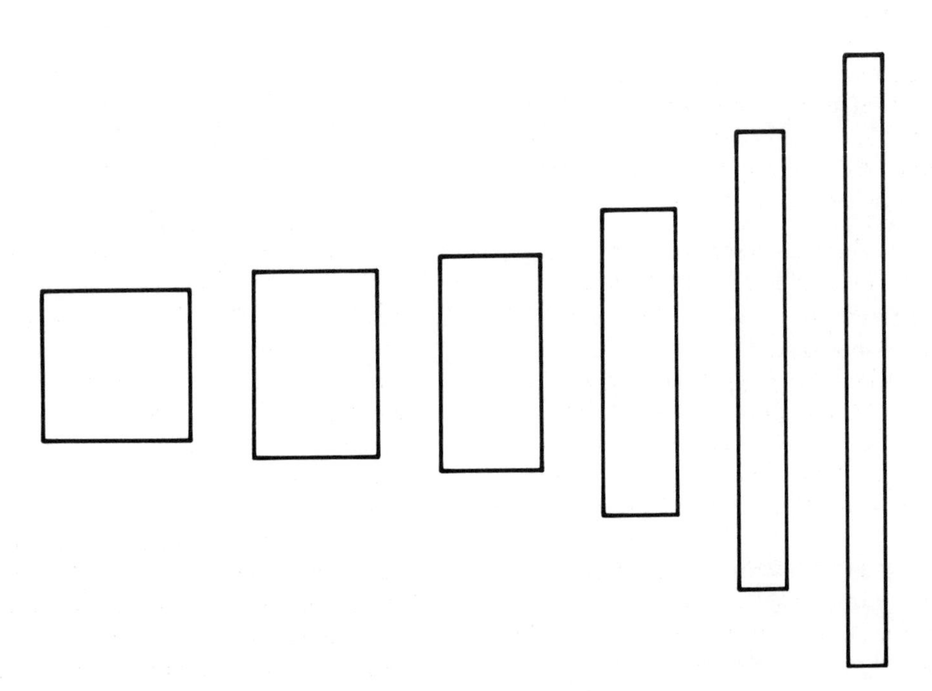

Do other similar sequences of fractions tend to limits, such as:

$$\frac{1}{1} \quad \frac{2}{1} \quad \frac{3}{2} \quad \frac{5}{3} \quad \ldots?$$

What kinds of rules might be used to create a sequence of fractions from an initial fraction?

D2 This problem is included because the answer is so surprising and elegant. Just as students in science are expected to, and do, appreciate phenomena which they are not expected to understand, and which scientists themselves do not always understand, so it is appropriate that students of mathematics should have the opportunity to marvel at phenomena which they cannot be expected to explain themselves.

A computer is useful for summing this series, which converges quite rapidly.

9 Powers and roots

Indices are a brilliant example of mathematical notation. The notation is simple, clear, and entirely does without the letters, names, or brackets that are part of other elementary notations such as sin x, $\cos(x+y)$ or $f(x)$.

Nevertheless, the mathematical notation of indices is only a convention, code to be translated, and a fact of mathematical life; it should not be regarded as a problem in itself, which is why there are no problems here on its basic meaning. If pupils forget what it means, they can be reminded, and pupils meeting the notation for the first time may well start a problem such as **A1** by writing out the powers as products.

Calculators have the usual advantages and disadvantages. They provide quick and good approximations to $\sqrt{2}$, for example, but many pupils will, and should, have a good enough idea of this to think about **B2** without using a calculator.

A computer can calculate 3^{20} exactly, but although this gives a vivid and valuable demonstration of the size of such numbers, it is the crudest way to answer the question.

On the other hand, being able to actually calculate powers and roots allows pupils to experience their size, and the power of the notation to represent numbers which it would otherwise be difficult, or impossible, to write down.

Problems **A1–5** and **B2** are expressed in terms of actual numbers, which makes them as easy as possible to tackle. Problems **B1**, **B3**, **C1** and **C2** have been deliberately expressed in more general terms. Algebra is not mentioned and is not essential for solutions, though it can be very useful, but the biggest difference is in the level of language. Not only is this grammatically more complex, but it is also more ambiguous, so pupils have to ask themselves not merely 'what does the English say?' but 'what does it mean mathematically?'

Some pupils will be aware that understanding the problem is the first part of the problem, and can face the harder language. For other pupils these same questions can be expressed in simpler language, for example in terms of specific numbers.

The roots are only square roots, with the exception of **C1**. It is appropriate that $\frac{1}{2}$ and $\frac{1}{3}$ should be used as indices in talking about square and cube roots, because pupils generally learn to appreciate new notations most easily if they use them as code first, and only later study their properties.

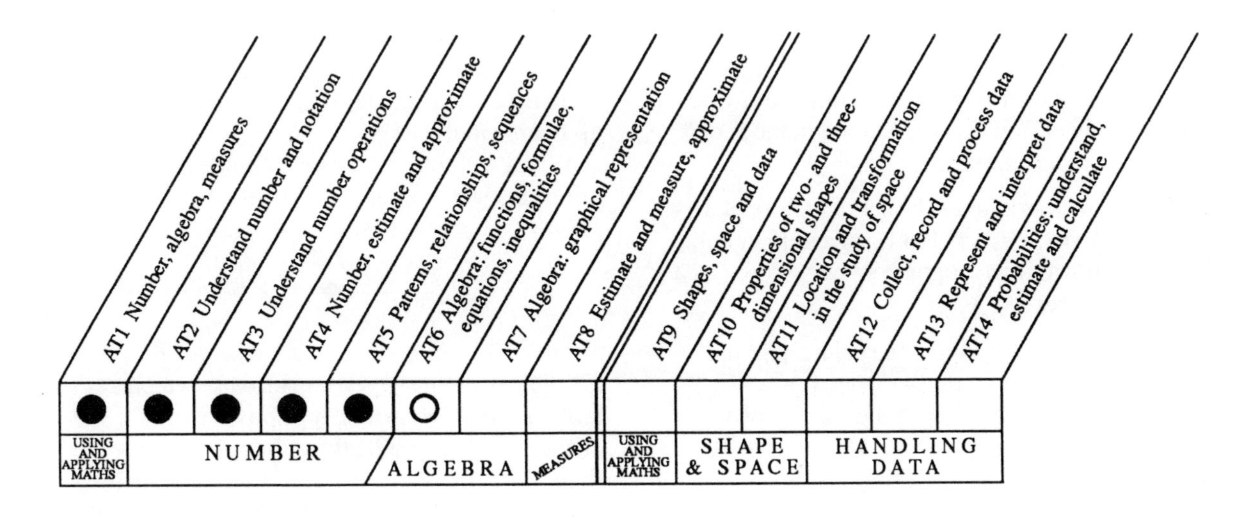

9 Powers and roots

Comments and suggestions

A1 How can a rough estimate be made mentally of the size of (say), 8^5?

A few bits of mathematical general knowledge are helpful in making estimates of powers, for example that 2^3 is rather less than 10 and 2^{10} only slightly more than 1000.

What is the connection between powers of 8 and powers of 2? (For pupils who have studied bases, how does this relate to the connection between binary and octal?)

A2 More generally, when is a^b greater than b^a? When is $a^b = b^a$?

Do you get a greater answer by raising a small number to a high power, or the same high number to the small power?

How close can such a pair of powers be? (In general, how close can two integral powers be?)

Small examples, such as $2^3 < 3^2$ or $3^4 > 4^3$ suggest there is no simple trend. A wider sample of cases may suggest a solution, provided it is safe to assume that the trend for moderately small integers continues for much larger integers. The very idea of such a continuity, and the qualification that it cannot always be assumed, is an important general concept, which problems such as this will help to develop.

A3 How do the prime factors of a power depend on the factors of the original number? What powers have fewest factors? How many factors does the product of two powers have?

How many factors, prime or composite, does a power have? Can you predict the number of factors of, for example, $3^4 5^2$?

A4 What property does the set of factors of a perfect square possess? How can a perfect cube or fourth power be recognised when written as a different power?

A5 This is the 19th Mersenne prime, and the largest known prime number when it was discovered in 1961 and also the first known prime number with more than 1000 digits. To find the number of digits exactly requires logarithms; how can the digits be counted approximately?

Apart from estimating the number of digits, how can the final digits of the number be calculated, without of course knowing the whole number?

B1 The form of the problem is chosen to be as dynamic as possible: the solver is given a choice – which way will he or she choose? Compare the use of the active verb *predict* as a dynamic alternative to the flat *calculate.*

Is there any special reason for adding three numbers, rather than two or four numbers? The general problem seems to be about the effects of adding first or squaring first. If adding first gives the largest answers for two numbers, would you expect the same result for three numbers?

Is squaring especially significant? Would cubing, or calculating any higher power, lead to the same conclusion?

An experimental investigation may suggest the more specific question: what is the pattern of differences between the two calculations? How much greater is one result than the other?

Pupils who are familiar with negative numbers might consider whether the same result holds if one or more of the numbers is negative.

B2 This is a kind of converse of the previous problem. Just as the squares of x increase at an increasing rate as x increases, so its positive square root increases more and more slowly.

The language of the problem, asking which is smaller, deliberately contrasts with the previous question, which asked for the larger.

B3 This question is asking, in effect, should you square the numbers first, to get the most out of the larger of the two numbers, or do you average them first, to bring the smaller number as it were up nearer to the larger?

When will the answer be: 'It makes no difference'?

C1 Experiment suggests an answer, which a table or graph confirms. If pupils are only accustomed to finding the cubes of integers, then both tables and graphs, but especially the latter, suggest that the cubes and cube roots of non-integers should be considered as well.

(How can cube roots be found on an ordinary calculator?)

How fast do fourth and higher powers or roots increase?

C2 The qualification 'large' is put in the question to get away from a feeble reliance on the small cubes that pupils may be familiar with. The pattern of differences between cubes can be studied as an integer sequence. How can the pattern of differences be continued without calculation of the original sequence of cubes?

What is the difference between N^3 and $(N+1)^3$?

This is another approach to the intuitive understanding of the sizes of powers and roots.

In general, how many digits do you expect in the product of, say, a four-digit number and a five-digit number? How many digits in the product of three three-digit numbers?

——— ◆ ———

D1 The googol was given its name by a nine-year-old child, the nephew of Edward Kasner (author with James Newman of *Mathematics and the Imagination*). It illustrates perfectly how in language we can very easily name mathematical objects which are hard to comprehend, and sometimes hard to find out about, or which do not even exist. (An example of the latter is an odd perfect number, an odd number which is equal to the sum of all its factors, excluding itself. If such a number exists then quite a lot is known about it, but no one knows if any exist.)

The same child invented the name googolplex for 10^{googol}. Even the intuitions of professional mathematicians fail before such gigantic numbers.

D2 There is an ambiguity in this number for anyone not familiar with the usual convention. Does it mean $9^{(9^9)}$ or $(9^9)^9$? The convention is that it means $9^{(9^9)}$. What is the difference between the two expressions?

9^{9^9} has the curious distinction of being the largest number which can be expressed in the decimal system with three digits and no extra symbols.

What other large numbers can be made with simple combinations of digits and symbols? What symbols are most effective in creating giant numbers? What symbols could be invented for this purpose?

(This is a serious question. Numbers appearing in some recent problems in combinatorics are too large to be represented in the usual notations, and so special notations must be invented before they can be written down or talked about.)

——— ◆ ———

E1 Pupils can use the index $\frac{1}{2}$ for square roots, with the advice that they will understand one day why this is done, and the suggestion that they might possibly realise for themselves, long before they try to understand the reasoning involved. As usual, the index $\frac{1}{2}$ is easier to appreciate than other fractional indices.

Before pupils study the indices for cubes and other roots, they need to have intuitive understanding of the size and variation of those roots.

E2 Such problems implicitly depend on the assumption that mathematics ought to make sense and that when you meet unusual combinations of symbols, or unusual ideas, you can decide what they do mean, or what they should mean, by asking yourself 'How can I make sense of this?'

This is not just a simple-minded approach, useful for nudging pupils into accepting what the textbook and the teacher knew all along: it is the approach that professional mathematicians take when they reach similar problems.

Pupils, having decided what 4^{-1} might mean, then need to also do what mathematicians do and ask themselves: 'Is this interpretation consistent? It makes sense in this particular example, but will it always make sense?' If the answer is yes, then they can safely go ahead and use it.

10 Rational and irrational numbers

Calculators and computers are very useful for many of these problems, though they are not essential, and can be misleading by providing shortcuts which lose information.

Rational and irrational numbers are distinguished on most syllabuses, but the distinction is not explored and pupils are not expected to develop any intuitive feeling for the differences between them. These problems explore the distinction, for three reasons.

They provide experience of fractions and decimals in a motivating context. There are many strong patterns to be observed, and explained, as well as some striking non-patterns.

There is something very mysterious about irrational numbers, much like the mystery of series and sequences. They go on for ever, without repeating, so in a sense you cannot 'know' them exactly, and yet you do 'know', for example, what $\sqrt{2}$ is – it is the number whose square is 2, or the diagonal of a unit square.

Decimals have their own properties, just as the integers do. There is a large difference between doing simple sums with decimals and appreciating their properties, though the latter will help the former indirectly by developing pupils intuitive feeling for decimals.

The stronger a pupil's intuitive feeling, for example, the harder it is to confuse 0.001 and 0.01, or not to realise that the second is no less than 10 times as great as the former.

Because these problems are largely to do with the properties of decimals from a pure mathematical point of view, there are only two problems concerned with errors and approximations.

The language of terminating, repeating and non-repeating decimals is essential. It is a bonus that the concepts involved are not limited to decimals alone. Repeated processes in general produce results which either repeat themselves, or terminate, or continue without doing either.

An example is the well-known and exceedingly difficult '$3n + 1$' problem, also known as Hailstone numbers or the Syracuse algorithm. Choose a number. If it is even, halve it; if it is odd, multiply by 3 and add 1. Repeat. What happens? Do all numbers terminate at 1? Do some of them enter a loop and go round and round for ever? Is the fate of some numbers to do neither? No one knows!

Only a couple of the problems actually use the word *rational*, and none of them uses *irrational*. As usual, pupils can learn the terms when they are useful to describe the pupils' experience, rather than being presented before that experience.

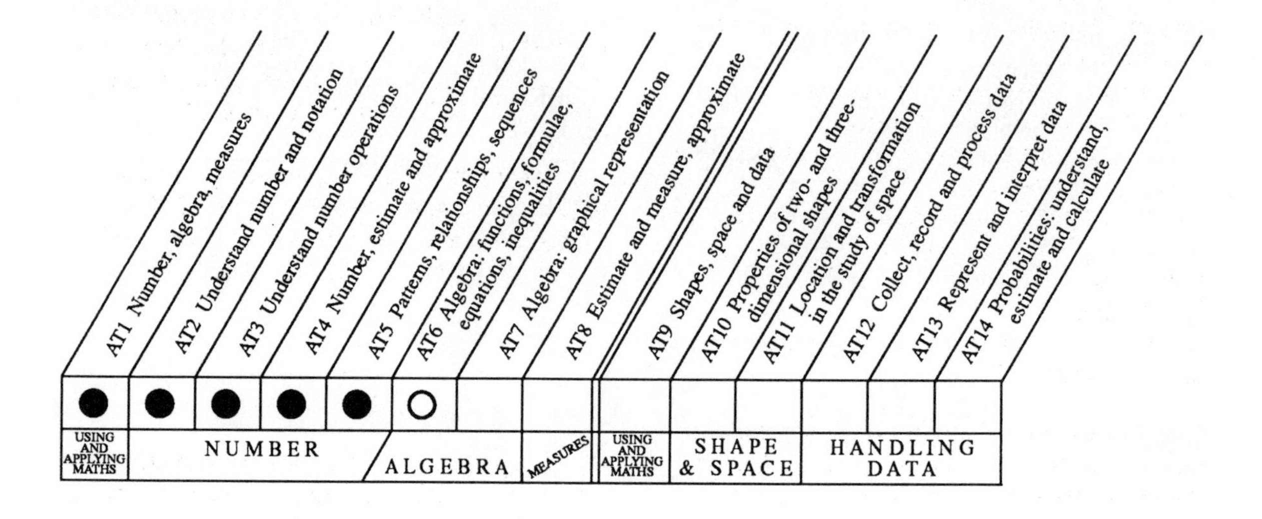

10 Rational and irrational numbers

Comments and suggestions

A1 Calculators may seem a valuable aid in such problems, but they will fail to show the pattern in the reciprocals of larger numbers, such as 17 and 19, and may suggest falsely to pupils that no pattern exists. This raises pertinent questions about the limitations of calculators. (There are ways of finding the reciprocals of larger numbers using an eight-figure calculator, but they depend on experience which pupils are unlikely to possess.)

The sample of reciprocals from $\frac{1}{2}$ to $\frac{1}{11}$ is quite a small sample. What happens when the sample is extended?

A2 Typically, the factors of the number n determine whether $1/n$ repeats or terminates. How can knowledge of its factors be used to predict the number of digits on a terminating decimal, or the length of the period of a repeating decimal?

A3 This problem suggests the idea that all repeating decimals are equal to some fraction, which is a much larger claim than that all fractions are equal to some repeating decimal. Can it really be true that a decimal point followed by any repeating sequence of digits – chosen at random if you wish – is equal to some fraction? The example of $\frac{1}{6}$, or $\frac{1}{12}$, suggests that even if the repeating portion does not start immediately, the decimal is still equal to some fraction!

Is there a one-to-one correspondence between decimals and fractions?

This repeating decimal may have appeared in a table of evidence for **A1**. If not, it can be roughly located by noticing that it is a little larger than $0.025 = \frac{1}{40}$. Alternatively, the reciprocal by calculator of 0.027 027 0 is 37.000 037.

What is the connection between the answer, and the number 27?

A4 The power and limitations of the calculator appear here also. Finding the reciprocal of the first few digits will suggest the answer if the fraction is a unit fraction, without suggesting any reason or rationale. If the fraction is not a unit fraction, the calculator will be helpless.

One idea is to consider a special case: the repeating decimals 0.010101010, 0.001001001001 ..., 0.000100010001 ..., and so on, by analogy with $0.111111 ... = \frac{1}{9}$.

Then, for example,
$0.68686868 ... = 68 \times 0.0101010 ...$.

A5 Why might not a fraction when expressed as a decimal produce an apparently random sequence of digits, like the decimal for π or $\sqrt{2}$?

The solution to this problem is only likely to be realised by pupils who have calculated reciprocals by division by hand, not on a calculator, and have spotted the pattern of remainders carried forward in the sum.

Each stage in a division sum can be thought of as a little algorithm for calculating the next decimal place in the answer. The whole sum is then an iteration – the repetition of the same routine again and again until the solution is reached. Iterations in general either go on for ever, come to a stop, or eventually go round in a circle.

A6 Any fraction certainly terminates or repeats, so a 'number' which does neither is no fraction.

It is certainly possible to write down a string of digits with no apparent pattern and imagine that it goes on for ever. This could be done with a die, for example. The result could certainly be marked approximately on the number line – but that only means that the number corresponding to the first 2 or 3 decimal digits could be located.

In what sense might the imaginary endless sequence of digits represent a number? Have pupils come across any 'proper' mathematical processes that create such sequences?

———◆———

B1 The pattern is easy to spot, but explaining it is much more difficult. It depends on thinking about the process of doing the necessary calculating by hand, and observing the remainders that are carried.

Are there similar patterns in the multiples of other repeating decimals, such as $\frac{1}{13}$ or $\frac{1}{17}$?

B2 The problem refers to the digits 142 857. These can be thought of either as a string of digits, in a particular order, or as a pair of numbers, 142 and 857, or as three two-digit numbers or as one number.

How do the properties of 142 857 compare with the properties of the decimal period of $\frac{1}{13}$ or $\frac{1}{17}$ or $\frac{1}{19}$?

B3 This is not just a trick, or an oddity. It points to the special significance of numbers such as 98 and 99, as well as to the connection between repeating decimals and geometrical series.

The answer is found quickly from the approximate solution given on a calculator, but this does not explain where the pattern 2, 4, 8, 16 comes from.

There is a similar pattern, arising for the same reason, at the start of $\frac{1}{8} = 0.125$, although that decimal terminates. The pattern 14, 28, 57 in the decimal for $\frac{1}{7}$ is another relative.

What similar patterns are found at the start, or end, of other decimal periods?

B4 Once again, the solution depends on the factors of the denominator, and denominators which are prime are the simplest.

Is it true that the periods associated with large denominators are necessarily longer than the periods associated with smaller denominators?

Given the lengths of the periods of, say, $\frac{1}{3}$ and $\frac{1}{7}$, is it possible to predict the period of $\frac{1}{21}$?

B5 This asks explicitly for the relationship which was noticed in the solution to **A3** and **A4**. It also goes somewhat further. Are all factors of a number of the form 999 999...999 the period of some decimal? Is every possible decimal period the factor of some number 999 999...999?

In which case is it true that whatever number you choose, provided it is not divisible by 2 or 5, there is a number of the form 999 999...999 which is a multiple of it?

B6 How can you even start to multiply an infinite decimal by 7? Depending on where you start, the final digit will vary, but all the other digits will be 9s.

Is there any difference between 1 and 0.999 999... ? Can you subtract one from the other?

Can any number be represented in more than one way in the decimal system? Does this matter?

———— ◆ ————

C1 This problem, and the next, both emphasise the difficulty of 'getting at' numbers which seem perfectly clearly defined, and which we can certainly talk about easily enough.

The traditional methods of calculating square roots by hand will calculate $\sqrt{3}$ as accurately as we wish, given time and effort. How accurate is $\sqrt{3}$ by calculator, as an approximation to the 'exact value' of $\sqrt{3}$?

What fractions are closest to $\sqrt{3}$? How close can a good approximation by fractions be to $\sqrt{3}$?

C2 This is the question that was not asked explicitly in problem **C1**.

Given any approximation, such as $3^2 = 9$, how can a better approximation be found?

Why is $3\frac{1}{6}$ a better approximation?

Why is $4\frac{3}{8}$ a much better approximation to $\sqrt{19}$ than 4?

C3 This is an example of another process, an iteration, which goes on for ever, but approaches a limit. The limit of this iteration also happens to be irrational.

What difference does it make if you start with a small or a large number? Would it make any difference if you started with a negative number?

What difference would it make if the conditions of the problem were changed slightly – for example, if the algorithm were

'Find the reciprocal, and add 3, then repeat...'

or

'Find the reciprocal, double it, add 2, then repeat...'

———— ◆ ————

D This simple question points to the subtleties that are only just below the surface we think of the sets of all rational or irrational numbers. How can such a straightforward question not have a straightforward answer?

11 Sharing and dividing

Fractions are very difficult for most pupils, with good reason, since here they first meet the advanced and abstract concept of equivalence, in fractions which are simultaneously equal but different.

Fractions are also very ambiguous. Their interpretation depends heavily on context. '$\frac{2}{3}$' could answer among other questions 'What is $\frac{1}{3}$ add $\frac{1}{3}$?', 'How many times does 3 go into 2?' or 'What is 2 divided into 3 equal parts?'

Pupils need the opportunity to tackle problems in terms of meaning before they start to develop formal methods which can all too easily be applied (often incorrectly, inevitably) with minimal understanding.

Historically, fractions have had at least four uses. They have appeared very simply as units of measure, as in halfpennies and farthings. Complicated fractions never appeared when they were used as units, not least because by, for example, dividing a foot into 12 inches, common fractions of 1 foot become integral numbers of inches.

Complicated operations with fractions did appear in gauging, measuring the contents of containers, a once-important practical problem that has largely disappeared due to standardisation and decimalisation.

Proportional division, such as sharing the profits of a business according to each partner's investment, was and still is important.

Finally, ideas of proportion and what used to be called the 'Rule of Three' must always be vital. In practical terms, if 25 litres of wine cost £64, how much will 32 litres cost? The concepts behind ratio and proportion are not only essential for all kinds of elementary calculations, but are also behind concepts such as density and pressure in science.

The problems in this section are all aimed at developing intuitive understanding rather than any formal skill. It is especially important that intuition is emphasised before skill because fractions are so easy to manipulate formally. What pupils find hard is to understand the reasoning behind the formal patterns.

It is too easy for pupils to spot patterns and latch on to them, very often pre-empting deeper understanding, which is only later revealed when they fail in practical situations.

Fractions can be interpreted statically, as objects, $\frac{1}{2}$, $\frac{2}{3}$, $4\frac{1}{7}$, or dynamically, as operators, $\frac{3}{4}$ of... $7\frac{3}{5}$ of/times.... The static interpretation is associated more with addition and subtraction, and the dynamic with multiplication and division, though this classification is not at all strict.

Integers can also be interpreted in these two ways, but the difference between the two approaches seems more confusing in more difficult contexts such as fractions and directed numbers. Some of the problems aim at a static interpretation, such as **C4**; others are aimed at the 'dynamic' interpretation, such as **D3**; others such as **A1** can be looked at either way.

The fractions appearing in the problems generally have small numerators and denominators. Almost all the problems can be made much harder by making the numbers larger, or by introducing fractions and decimals into their statements.

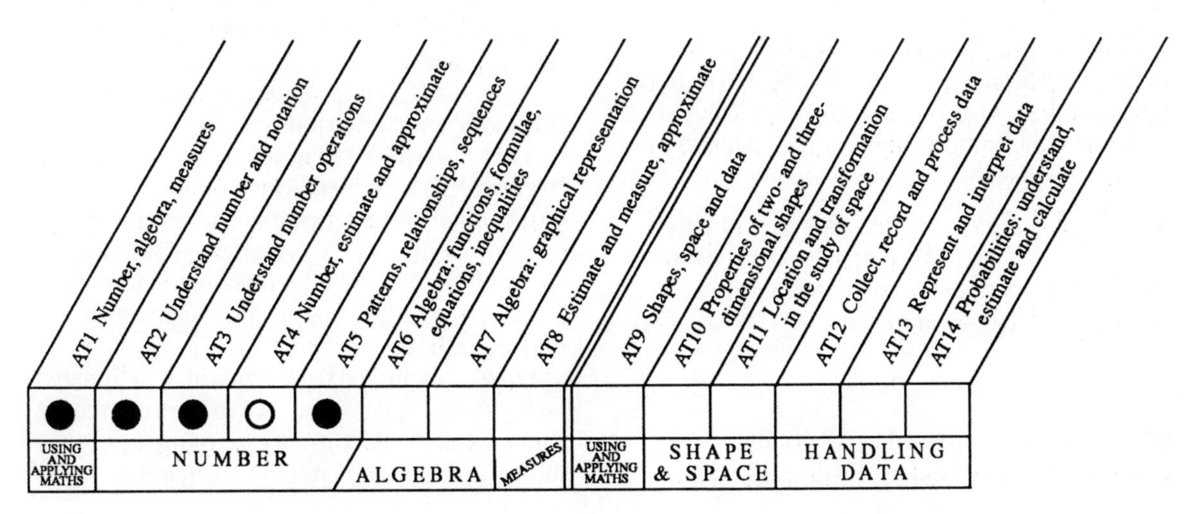

11 Sharing and dividing

Comments and suggestions

A1 The symmetries of the rectangle and the properties of the number 12 are both important in such problems. Taken together, they make this problem richer and more interesting.

Is *equal* ambiguous? Does it, or might it, mean equal in area or equal in shape and area?

What different numbers of parts would make the problem more difficult? Would starting with a square make the problem any easier?

Which divisions of a rectangle can be performed accurately by using its symmetries alone, and which can only be done accurately with the use of a ruler?

A2 How does the 'three-ness' of the triangle make it more or less easy to divide into, say, 3 or 5 parts?

How much easier is the problem if the triangle is equilateral? How much easier does it seem if the triangle is equilateral?

Once again, the problem might mean 'equal in area' or equal in area and shape, that is, congruent'.

Does each part have to be in one piece, or could 'someone get his or her share' in several small pieces?

A3 What could 'one half of a part' mean? A perfect analogy would be sharing something between 3 adults and 1 child, so that the child gets a half portion.
$3\frac{1}{2}$ is a relatively simple number. What about $4\frac{3}{5}$ parts? What might $\sqrt{2}$ parts mean? What might π parts mean? In what sense can the circumference of a circle be divided into π parts, each equal in length to its diameter?

A4 Once again the number of parts, 8, is chosen to relate to the symmetry of the cube. Into what other numbers of parts can a cube be easily divided? How could a cube be divided into 7 or 9 parts?

A5 Dividing the rectangular bars individually into 5 pieces is easy. What happens if there are 3 bars? What possibilities are there apart from dividing each bar separately?

———◆———

B1 This is closely related to **A3**. Many problems can be solved by dividing the whole into a suitably larger number of parts. This in turn relates to the convenience of dividing a circle into 360 parts, or a foot into 12 parts, and the relative inconvenience of dividing a metre into 100 parts.

Into how many different numbers of parts can £10 be conveniently divided?

B2 Larger numbers frighten off some pupils. £32 000 makes the problem seem much harder than a problem about £10. Why is it impossible for the sharing out to be performed exactly? How can this be deduced quickly from the figure of £32 000?

Why must it be possible to share exactly, if one of the three recipients agrees to go up to 2p short, not more?

B3 Like many problems in fractions this depends on the properties of the number 720, specifically its factors.

In what sense is 720 a 'round' number? How might the 'roundness' of a number be measured? Are some numbers 'round' for one particular purpose, but not 'round' for a different purpose?

———◆———

C1 This problem takes aim at the pupils' intuitive understanding of fractions. How many fractions are there? Where are they? How close together are they? Are there any spaces on the number line which do not contain fractions? How small can the difference between two fractions be? What fractions are very close to $\frac{2}{3}$? What 'simple' fractions with small denominators are close to $\frac{2}{3}$? How close together can 'simple' fractions be?

As so often, such very pure questions lead to surprising and mysterious results.

They also exercise pupils' ability to compare fractions, perhaps by marking them on a number line, more efficiently and informatively by subtraction, or in many cases quickly and easily by arguing, for example, that $\frac{4}{7}$ is less than $\frac{4}{6}$ and therefore less than $\frac{2}{3}$, and outside the range of the question.

How can you find fractions which definitely lie between $\frac{2}{3}$ and $\frac{3}{4}$? One way to 'average the two fractions' is to take the average of their numerators and denominators separately $2\frac{1}{2}/3\frac{1}{2}$

or $\frac{5}{7}$. This is also the result of simply adding the numerators and denominators. Why does this fraction necessarily lie between the two initial fractions?

How could this fact be used to find more and more fractions between them?

(Calculators can be an aid here, but as usual they hide rather than reveal any patterns that pupils might otherwise spot.)

C2 Should equivalent fractions such as $\frac{3}{4}$ and $\frac{6}{8}$ be counted separately? The mathematical approach would be to consider both possibilities and decide from the results which is most profitable, or most aesthetically pleasing, rather than making a snap decision before the evidence has been considered.

What properties, if any, does the sequence of fractions with denominators up to and including 10 possess, when they are arranged in order of size?

Is it possible to predict the number of fractions with denominators up to a certain number?

C3 Such pure problems are suggested by one of the strategies for solving **A5**. How many sets of three reciprocals can sum to 1, bearing in mind that at least one must exceed $\frac{1}{3}$ and one must be less than $\frac{1}{3}$?

How many sets of 4 reciprocals sum to 1?

How can one set be used to create another set?

How many infinite sequences of reciprocals sum to 1?

C4 It is a convention, nothing more, that the set of fractions equivalent to $\frac{7}{8}$ includes only fractions whose numerators and denominators are integers.

In practice, fractions such as $3\frac{1}{2}/4$ occur frequently, (and even more often in the form of proportions, in which this convention has never applied.)

Should $24\frac{1}{2}/28$ or $9\frac{1}{3}/10\frac{2}{3}$ be counted as a fractions equivalent to $\frac{7}{8}$?

Should $-\frac{7}{8}$ also be counted?

C5 The emphasis is placed on meaning, because of the convention that fractions have integral numerators and denominators. As so often, we can go safely outside the convention provided we think carefully about what we are doing – about what such fractions mean.

Fractions such as $\frac{3}{8}$ can be interpreted in different ways. Which of these interpretations carry over to $3\frac{1}{2}/8$?

C6 Compare problem **A3**. This is harder to understand than **C5**, where the fraction appears in the numerator. How can anything be divided into $7\frac{1}{4}$ parts? That might seem impossible at first sight. On the other hand, as a ratio it is not so difficult to understand. Surely any number can be compared to any number in size?

Does it answer the question, 'How many times will $7\frac{1}{4}$ go into 1?'

D1 What does it mean, or what could it mean, to talk of $\frac{1}{3}$ as a fraction of $\frac{1}{2}$? Does it help to think instead of $\frac{2}{6}$ as a fraction of $\frac{3}{6}$ or $\frac{4}{12}$ as a fraction of $\frac{6}{12}$? Does it help to think geometrically, by thinking of the positions of $\frac{1}{2}$ and $\frac{1}{3}$ on a number line?

Can any fraction be thought of as a fraction of any other given fraction? For example, is it natural and inevitable that $\frac{5}{14}$ is some fraction of $\frac{7}{9}$?

D2 What does, or might, the little word *of* mean in this problem?

Is it easier to think of two-thirds of $\frac{12}{15}$? Is the solution then intuitively more obvious – and does this way of thinking also serve to explain, with an intermediate argument, why the word 'of' in this situation triggers *multiplication* of the two fractions?

What does two thirds of four fifths mean in geometrical terms?

D3 The relatively large 14 and 10 are chosen deliberately, so that they contrast with the small fractions $\frac{1}{3}$ and $\frac{1}{2}$. How does the difference between these two products depend on the size of 14 and 10?

D4 Such questions can be thought of in rough and approximate terms by thinking of $3\frac{1}{2}$ and $12\frac{1}{4}$ on the number line, or even by mentally comparing 3 and 12, and realising that the answer is rather less than 4. This is what a joiner would do when deciding how many $3\frac{1}{2}$ lengths can be cut from a $12\frac{1}{4}$ piece of strip, and roughly what a shopper would do with £12.25 to spend, and each item costing £3.50. In such cases, the answer 3 times, or 3 times remainder $1\frac{3}{4}$ is not only adequate, it is perfect.

In other circumstances, for example in physics or chemistry, it is probably quite inadequate. But what might an *exact* solution be like? Is it possible to even answer such a question exactly?

12 Proportion

All problems in proportion can be interpreted in terms of fractions, but that is not to say that fractions and proportions are indistinguishable. A fraction is a 'thing', a number. It can be marked on a number line, and a fraction of, say, a piece of cheese or a rectangle is about as substantial as the original object.

A proportion, in contrast, is a relationship, and so thinking in terms of proportions promotes a more dynamic way of thinking and of looking at problems. This dynamic approach often matches pupils' everyday intuitions about relative sizes and quantitites in a way that the language and concepts of fractions do not.

This is not to say that it is 'better' to think in terms of proportions rather than fractions. Rather, both ways of looking are valuable and, on occasion, essential. Pupils who are only able to think from one of these complementary points of view will be disadvantaged compared to pupils who have both perspectives at their command. The language and concepts of proportion are especially useful when thinking about such problems as **A1–3**.

Some of these problems use the ':' notation for proportion. This closely matches the everyday use of English in sentences such as '20 is to 10 as 8 is to 4', and it focusses on the relationship between the two numbers without the intervention of a third number, a fraction, which 'equals' or 'stands for' that relationship.

This use of different notations is not confusing to pupils, provided they are discussed and talked about. Rather, the use of the most appropriate notation and concepts in a particular situation makes understanding easier.

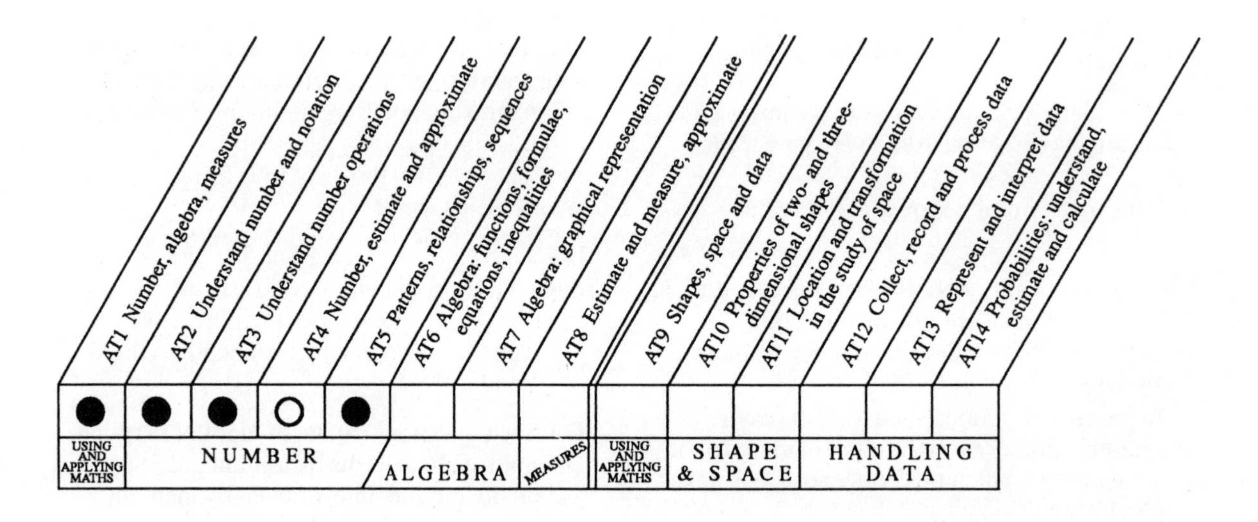

12 Proportion

Comments and suggestions

A1 There are at least two obvious solutions to this problem for pupils who have not met such problems before and who do not appreciate the significance of 'in proportion'. 10 is to 20 and 7 is to 17, where the difference is preserved, and 10 is to 20 as 7 is to 14, where the ratio is preserved.

A2 In **A1**, the very simple fact that 20 is double 10 prompts the answer 14, even in pupils who are not familiar with the word 'proportion'.

In this problem, the proportion is more complicated, and needs more thought, and pupils need to understand 'in proportion' more clearly. Such problems make good discussion points. Accepting that it is the ratio of the numbers which are important, how can 25 be compared to 10? Or how can 80 be compared to 10?

Can the problem be solved by comparing 10 with either 25 or 80?

A3 When neither comparison is quite as simple as it might be, the question is prompted: how can proportions be solved in general? By what algorithm, which will work however complicated the numbers are?

A4 The idea of a proportion appears in its 'purest' form when all context is missing. The idea that numbers in proportions could be integers, or fractions, or decimals, (or even irrational!) is not necessarily obvious to pupils.

The basic property of proportions, that a proportion does not change when both terms are multiplied or divided by the same factor, can be illustrated by diagrams such as this:

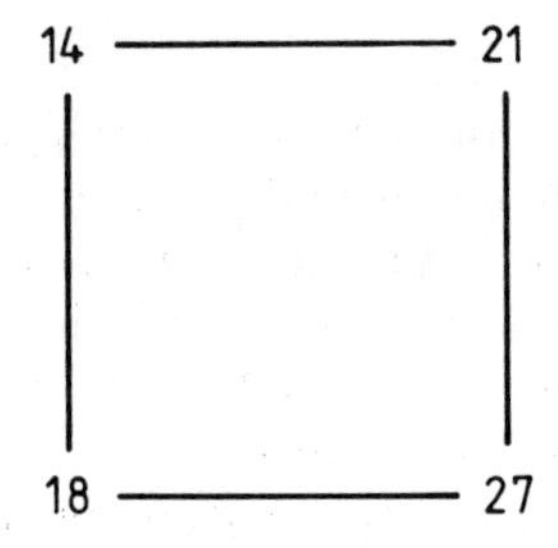

It can also be illustrated by breaking up proportions such as $1\frac{1}{2}$:4 in this problem, like this:

$$\frac{1}{2}+\frac{1}{2}+\frac{1}{2} : \frac{1}{2}+\frac{1}{2}+\frac{1}{2}+\frac{1}{2}+\frac{1}{2}+\frac{1}{2}+\frac{1}{2}+\frac{1}{2}$$

Now it is intuitively clearer that this is the same proportion as 3:8, which can be obtained immeditately by doubling the original terms.

A5 Equivalent proportions can be illustrated geometrically, in many ways. This is a simple possibility:

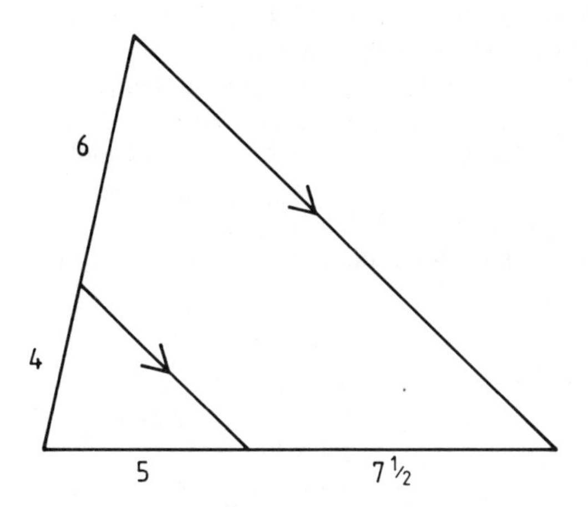

$4:6 = 5:7\frac{1}{2}$ and $4:5 = 6:7\frac{1}{2}$

The possibility of drawing as many lines parallel to the original number line as we choose, suggests how many proportions are equal to a given proportion, as well as suggesting that, for example, the '12' in this problem could be replaced by any number at all – including irrationals such as $\sqrt{2}$ or π – and the proportion could still be completed.

A6 Why are decimals in a proportion equivalent to proportions with fractions? How can they explicitly be turned into proportions in fractions only?

A7 A fraction such as $\frac{5}{8}$ is conventionally considered to be equivalent to $\frac{10}{16}, \frac{15}{24}, \ldots$ and so on. Much greater variety is conventionally allowed to proportions.

Can the proportion $4.2:5.0 = N:\_\_$ be completed whatever number appears as N?

A8 This problem invites the general query: is any proportion equal to a proportion between integers? The Greeks, who thought naturally in terms of proportions between geometrical lines or areas, discovered the answer. $\sqrt{2}$ cannot be so represented. Neither can any square root of an integer that is not a perfect square, or numbers such as π or e.

However, all proportions between fractions can be.

What is the general algorithm for turning such a proportion between fractions into one between integers?

B1 This method is actually used by biologists to estimate populations. It is striking that it allows an estimate of a population which cannot be counted whole, through there are obvious sources of error. How do you know that the captured animals are a random sample each time?

The technique can be simulated, without these crude errors, by using pebbles or marbles in a closed bag which is shaken thoroughly between each sample withdrawal. How accurate is this method?

B2 In a simple chain of gear wheels their circumferences are locked together, and therefore it is the number of teeth which pass a fixed point which is important.

In this gear chain, given the rate of rotation of the left-hand gear, that of the fourth gear can be deduced without knowing the sizes of the intermediate gears:

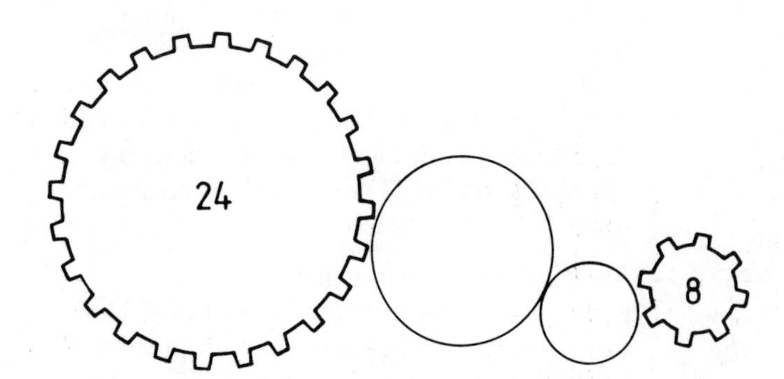

What difference does it make in the problem that the middle two gears are on the same axle?

B3 What is the largest model, keeping the sizes and relative distances in proportion, which would fit inside the classroom? Or inside a shoe box?

How many figures in the given dimensions of the earth, moon and sun are significant on such a small scale?

B4 What everyday objects can be covered more or less exactly by, for example, a penny held at arm's length? A 10p piece held in the same position?

How could a simple instrument be designed to discover the relative sizes of objects which are about the same distance away, but are too far away to be reached?

C1 The proportion expresses the proportion between 20 and 15 and 20 and 28, despite the unsymmetrical positions of 15 and 28. A diagram such as the one in **A4** is quite impractical, but is does illustrate the possible relationships more symmetrically.

C2 This is a discussion question, rather than a problem to be solved.

Why are some problems easier to think about when written as proportions than as fractions, and conversely?

Where do mathematical notations come from? How many mathematical notations are as simple as they could be? What are the simplest and most brilliant notations?

C3 The problem of designing such an algorithm appropriately follows the pupils' wide experience of solving problems in proportions, and matching problems expressed in terms of fractions, by common sense, and by elementary and very simple reasoning with small and simple numbers.

On the basis of such wide experience, involving different mental processes for solving problems such as **A1**, **A4** and **A6**, pupils develop a strong intuitive feeling for what they are doing, which will aid them when they have to handle more complicated numbers.

An actual algorithm can be thought of as a summary of the most general ideas they have used. Forgetting about the little tricks, the relationships that you can often spot, what are the most general and powerful arguments that always get to the solution – even if sometimes they go a long way round?

One powerful way to conceptualise the problem of describing an algorithm is to imagine that you are actually instructing a computer, or telling a partner how to complete a proportion without knowing yourself what numbers are involved.

———◆———

D1 Problems of this type have always been important in practical applications. How much more difficult would the problem be if the division was into 24 parts of copper and 13 parts of zinc? How could the problem be solved if 297 cm^3 of liquid are composed of 102 parts of sugar solution to 41.7 parts of water?

D2 This is equivalent to the question in Unit 11 asking for fractions between $\frac{2}{3}$ and $\frac{3}{4}$. What fractions, with simple small denominators, are close to a given fraction?

How can all such good approximations be found? How can they be searched for informally? What proportions are hardest to approximate with another?

13 Measuring in two directions

By focussing on measuring in two directions, these problems give a geometrical flavour to elementary ideas of negative numbers. To some pupils these ideas may already be familiar – for example, if they have met negative numbers as coordinates on maps or in playing board games.

Ordinary integers can either be interpreted statically, as things, or dynamically, as operators – for example, when a debt is multiplied by 2. In ordinary arithmetic and mathematics this difference is naturally elided most of the time for the same reason that no distinctions are made between '–' as a sign and as an operation: if you look at them deeply enough, they are essentially the same.

This is by no means obvious to pupils. Too often, only one aspect is presented, or both aspects are presented but only implicitly and indiscriminately, so pupils never face up to and appreciate the two possible viewpoints.

Geometrical interpretations naturally lend themselves to either a static or a dynamic interpretation, depending on whether numbers refer to static positions, or to dynamic movement.

This idea can be and should be discussed in the present context, so that pupils realise that there are two distinct ways of looking at so many expressions, and that these different ways are not contradictory, and that either interpretation is possible.

It is especially relevant because of the vast difference in difficulty between the appreciation of addition and subtraction of directed numbers, or their multiplication or division by ordinary numbers, and the use of directed numbers as operators themselves.

Primary pupils can understand R5 + L3 = R2, or 4 × R6 = R24, especially if they are interpreted dynamically, as movements. Most secondary pupils have great difficulty in understanding what it means to multiply any number positive or negative, by a negative number.

There is no mention in these problems of the usual notations for positive and negative numbers, though some pupils may have come across them. They are not necessary to solve the problems, and indeed it is much better for pupils to invent their own notations, and to discuss what they are doing, than to latch on to the standard potentially confusing notation before they are ready for it.

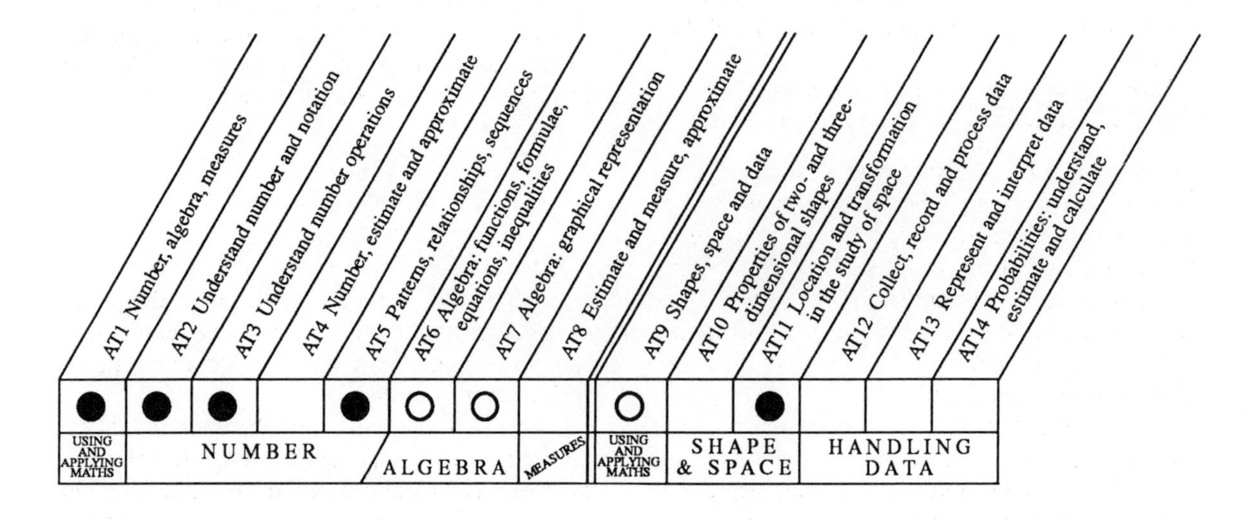

13 Measuring in two directions

Comments and suggestions

A1 The solution to this problem does not have to be the usual conception of negative numbers. This is especially so since the problem only refers to counting, not to any other operation.

Some pupils may well suggest starting to count upwards again. This is not an unreasonable answer, though it does raise the problem of distinguishing between the two 'ones'. How are such numbers distinguished by banks, or how were they distinguished before computers arrived? How are heights above and below sea level shown on maps?

A2 Different contexts may suggest different solutions. Distances along a line which happens to be drawn horizontally might naturally suggest 'right' and 'left', and their abbreviations as suitable distinguishing marks.

Note that as in **A1**, there is no suggestion here of operating with such marked numbers. The problem merely refers to measurement. However, pupils may well think of adding such numbers, or performing other operations with them.

A3 Like a line with a beginning but no end, a graph which includes only one quadrant is, as it were, unsymmetrical, bounded in two directions but not in two others, apart from the edge of the paper.

How could all the points in space be described by using three axes?

A4 Temperature is an instructive example of scientific measurement, because no natural zero was perceived by early scientists who therefore chose rather arbitrary zeros, such as the freezing point of water. The scientists concept of absolute zero is still unnatural in everyday life.

How are speeds and accelerations in different directions distinguished?

It is notable that in elementary scientific measurement, negative numbers are not used as multipliers. Positive and negative temperatures may be multiplied by natural numbers, and added or subtracted, but that is all. '5 below zero' multiplied by '7 below zero' does not mean anything. (However, '12 below zero' divided by '3 below zero' does have a value, the ordinary number 4, which is not a temperature, and a meaning: the first temperature is 4 times as great as the second, in magnitude.)

◆

B1–B5 The point of these problems is that positive and negative numbers are not used. Pupils have to think about them in other terms, red and black (an excellent example of the use of colour in mathematics as a distinguishing mark) or debit and credit.

By thinking in such terms, which are also important in everyday life, pupils can become familiar with what at first meeting can seem very strange situations.

If pupils can examine actual bank statements, or take part in a game or simulation in which they have to handle credits and debits, so much the better. Notice once again that only operations with natural numbers on debits and credits are involved here. It makes sense to add or subtract credits and debits, or to multiply or divide them by a natural number. It makes no sense to multiply a debit of £5 by a debit of £7, and the answer is certainly not a credit of £35! (This is stated flatly here, because asking pupils whether such an operation is possible is likely to confuse rather than enlighten them.)

◆

C1 Problems **C1** and **C2** are visual analogues of the number patterns which are commonly used to informally motivate the rules for operating with negative numbers.

The visual suggestion that the lines might be continued to the left is quite strong, though not as strong as the pattern of the parabolic graph in the next unit. The straight line is, as it were, too simple, and pupils are too accustomed to thinking of lines as naturally having at least one end.

If the line is continued to the left, will the points on it naturally fit the same equation? Or might they fit a different equation? Perhaps the signs in the equation will have to change?

Can all lines be extended across the vertical axis? Can lines which cut the horizontal axis be extended below that axis?

C2 Only addition of numbers is involved, matching the addition of debits and credits in earlier problems.

The question similar to problem **D1** also naturally arises: how do you compare a number which is large and less than zero, with a small ordinary number?

Another query: how can you show on a graph, pairs of numbers whose sum is less than some number which is itself less than zero?

D1 When negative numbers still mystified leading mathematicians, it was argued that if they really existed then $1/-1 = -1/1$ would mean that a larger number divided by a lesser number equalled the same lesser number divided by the same larger number, which appeared to be manifestly absurd. The apparent paradox was resolved by the realisation that negative numbers required two ideas of comparison. 1000 less than zero is indeed less than 0.0001, in one sense, but in the sense of magnitude, it is greater.

A scientific example is an electric shock which is equally dangerous whether the current is positive or negative, in contrast to the small shock of touching torch battery terminals with the tongue.

D2 Such comparisons are maximally confusing: 1000 less than zero is clearly large, in some sense, and 0.001 is just as obviously a small number in some sense.

It is a matter of interpretation. What is being talked about? In what real life circumstances might 1000 less than nothing nevertheless be less than 0.001? In what circumstances might it obviously be greater?

E R and L are as reasonable a way as any for distinguishing measurements in two directions, whether they are thought of as static descriptions of position, or as dynamic descriptions of movements to a position.

What other operations might be done with them? How could they be subtracted? How could they be multiplied by ordinary numbers? Might you possibly multiply or divide R and L numbers by each other? This last question is typical of the kinds of questions that mathematicians ask themselves in similar circumstances. Mathematicians often have to wrestle with such problems, so it is no surprise if pupils find such a question mysterious, and unanswerable, at the moment.

In the context of problem **A3**, how might positions, or movements, in a plane by added and subtracted? In three dimensions?

14 Negative numbers

There are two ways to interpret the relationship between negative numbers and ordinary numbers. One is to think of the ordinary numbers as unchanged, and the negative numbers joined to them, as an extension which allows quantities to be less than zero, provides many equations with solutions where they would otherwise have none, and so on.

In many ways this is the simplest interpretation, and is the interpretation chosen here. The negative numbers can be shown by a suitable notation which need bear no confusing resemblance to the minus sign. The bar over the top as used for logarithms is clear and easy to write, and can be read as 'bar...' or 'negative...' rather than as 'minus...' which invites confusion with the operation '...minus...'. The term 'positive' is then used as an alternative term for ordinary numbers, for purposes of emphasis, or to distinguish the positive values of an expression from the negative.

The usual problems then arise of operating with ordinary and 'bar' numbers together, and of their use in representing, for example, measurements in different directions along a line.

The alternative interpretation is to start with a new idea, of positive and negative numbers, and with a new notation for each, such as the common ^{+}x and ^{-}x. Mathematically it is equivalent to building a new structure, part of which is isomorphic with the system of ordinary number, rather than simply extending the latter.

Whichever interpretation is chosen, the same variety of uses of ordinary/positive and negative numbers have to be accounted for, and the same variety of levels of difficulty is present.

Easiest are calculations in which the negative/directed numbers are, as it were, the objects, which are operated on by an ordinary number. Thinking in terms of one number 'operating' on another is an active and natural way of thinking which is especially appropriate when the number operated upon is clearly a 'something' for example a distance along a line, or a bank balance.

About as simple are additions of ordinary/positive and negative numbers, as long as they are interpreted as, for example, movements along a line.

Next come addition and subtraction of ordinary/positive and negative numbers including the subtraction of negative numbers. Finally come calculations in which negative/directed numbers are operating on other negative/directed numbers. The very meaning of $\overline{5} \times \overline{6}$ is a moot point which is much harder to interpret than the meaning of $10 - \overline{7}$.

As in so many elementary mathematical situations, there is a danger that some pupils will pick up the 'pattern of use' of negative numbers without understanding what the pattern means or why it is necessary for practical applications. This is perhaps the more likely because the actual calculations are so simple.

To counteract this tendency it is necessary to focus closely on the meaning of what is happening, which is why many of these problems invite, or require, discussion and argument. Throughout this topic, calculation tends to be relatively trivial, and meaning is everything. Needless to say, it is meaning which so many pupils have traditionally not taken with them.

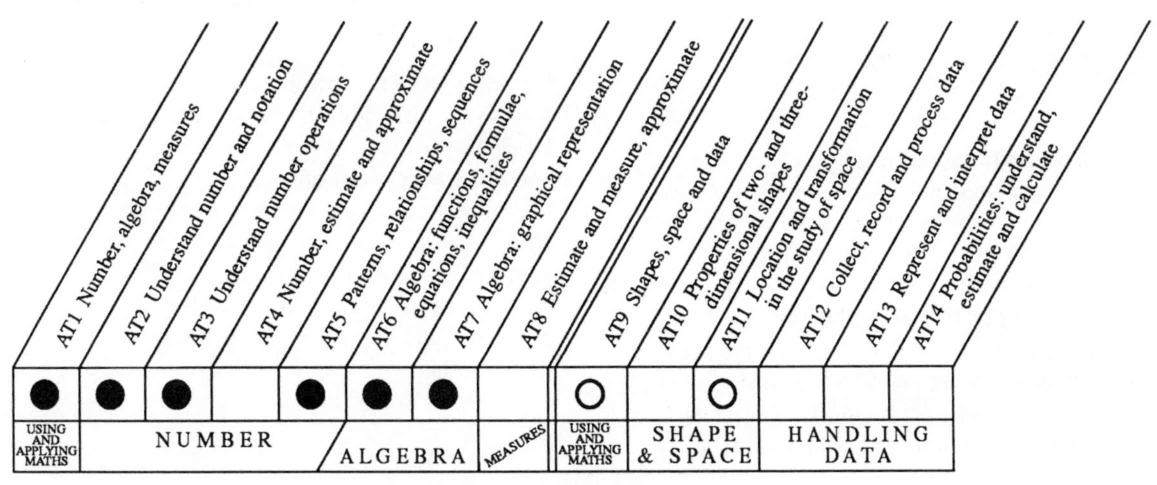

14 Negative numbers

Comments and suggestions

A1 'Double' and 'treble' are deliberately written in words, because they will be easier for pupils to intuitively understand than '2 ×' or '3 ×'. The qualification 'possible' is also used deliberately to emphasise that this kind of multiplication is not logically forced. The question 'Is it possible…?' then invites pupils to consider possibilities.

This problem queries whether a particular kind of multiplication,

ordinary number × negative number,

can be switched round. In what everyday or scientific circumstances does it make sense to multiply by a negative quantity? When is it nonsensical to do so?

The questions in the last two sentences are quite different. Many pupils may guess that $\overline{9} \times 4 = \overline{36}$, but what does it mean?

A2 As in **A1** there are two possible approaches, which are complementary. One rests on the extremely 'pure' idea that operations with negative/directed numbers should be consistent with operations with ordinary numbers. The other rests on finding suitable interpretations which force the usual conclusion if negative/directed numbers are to fit the interpretation.

Both these questions are for discussion, not mere 'answering'. It is especially important that pupils should appreciate the subtlety, difficulty and elegance of negative numbers. If they find negative numbers difficult to grasp, they will then be reassured that this is not merely their own stupidity.

◆

B1 This question does not use terms such as 'commutative' and 'associative' or 'distributive' because such terms are of little value until the occasion arises when the distinctions they embody become important.

They might be introduced while informally discussing a subject such as this, in which facts about ordinary numbers which are usually taken for granted become problematic as soon as negative/directed numbers are introduced.

B2 How will they compare in different interpretations? If numbers stand for movements in two directions, addition represents a combination of movements, and subtraction the difference between two movements?

If they are sums of money, credits and debits, what will + and − mean?

B3 Such patterns represent in a simple visual manner the assumption that whatever works with ordinary numbers, will continue to work with negative numbers also.

Many of the well known number patterns can be used for the same purpose, and pupils can also make up their own.

◆

C1 Such graphs of functions or formulae are effectively number patterns, waiting to be extended using negative numbers.

Compare the function and its graph for the volume of a gas at different temperatures ('the meaning of formulae' problem **A3**).

C2 Whether pupils are or are not familiar with the parabolic shape of this graph, it is natural to wonder how it can be 'completed' by adding values to the left, for which x is negative.

Can all graphs of quadratic expressions be continued across the vertical axis? What about the special case $y = x^2$ which, for positive x only, is not quite so obviously parabolic in shape as the problem equation?

What about graphs of more complicated expressions, such as x^3, or

$x^4 - 3x^3 + 2x - 8$?

C3 How does the graph behave as one of x and y gets very large and the other gets very small? Does it ever reach either of the axes? Is it, as it appears to be, symmetrical? About what axis?

This problem involves a possible confusion between 'less than' in the sense of absolute value and the 'less than' in arithmetical value. In a practical application it might well be the absolute values of x and y which were significant.

◆

D1 What can be said about two numbers whose quotient is 12? Or whose sum or difference is 12? The latter two bits of information say nothing about either of the number, illustrating one difference between addition/subtraction and multiplication/division.

D2 The same symmetry as in **D1** points to the existence of two square roots of any positive number.

This means that even a simple quadratic equation like $x^2 = 9$ has two roots, rather than the one positive root. Since equations such as $x^2 + x - 5$, already referred to, have two roots when extended to negative values of x, this might suggest that it is 'natural' for most quadratics to have two roots. What about the exceptions? What about the numbers of roots of cubic equations?

D3 No number, in the context of ordinary and negative numbers. However, if mathematicians can invent negative numbers, perhaps they can invent yet another kind of number which will provide square roots of $\overline{9}$?

Indeed they can. The resulting complex numbers provide an interesting topic for investigation by pupils who are confident 'algebraists'. The ease with which they can be manipulated formally, in contrast to the difficulty of deciding what other meanings they might have, matches perfectly the experience of using negative numbers.

———◆———

E It is an important fact that mathematics books not written to teach negative numbers use a '−' sign instead of a bar or a prefixed superscript '−'. Why do they do this?

What confusions are likely to arise? Some pupils are likely have picked up from somewhere or other the tag, 'two minuses make a plus'. What is the difference in meaning between the two '−' signs in $100 - -8$?

15 Errors and approximations

The subjects of approximations, estimations and errors are mathematically identical, but have different emotional connotations. Approximation and estimation are emotionally neutral, or even have negative associations for some pupils; estimation is what you are supposed to do to ensure that your solution to a problem is not absurd, and it can therefore be regarded as something of a chore.

Errors, in contrast, have very strong emotional connotations, not least because the word is associated with mistakes. While it may be unpleasant to face up to your own mistakes, tackling mistakes presented by someone else is a much more enjoyable and vivid experience, such is human nature.

The idea that errors are inevitable, and can grow, or multiply, or be reduced to a minimum is likewise a vivid idea.

It is also an important topic because the manner in which errors are affected by addition, subtraction, multiplication and division is closely related to concepts of elementary algebra, especially to products such as $(a+b)(c+d)$. The errors that rapidly accumulate when a number is raised to a power are similarly related to algebraic expansions of higher degree.

The conventional notations for errors and approximations are not introduced in these problems, precisely because they are conventional, and can be introduced as and when pupils need them.

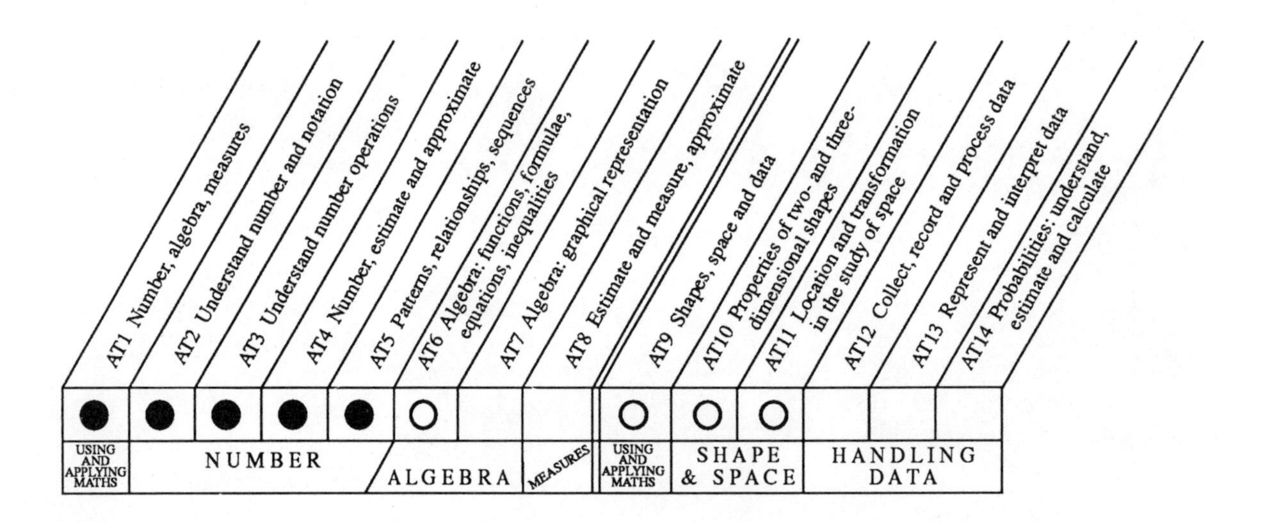

15 Errors and approximations

Comments and suggestions

A1 In practical problems the maximum error is often at least as important as the average error. Fortunately, errors increase relatively slowly when accumulated by addition only, and when the errors can go either way the accumulated error could be small or even zero.

How far out are the four pieces likely to be, on average, when fitted end to end? If each piece is in fact either 1 mm short or 1 mm too long, what is the chance that they will 1 mm out?

How do the likely errors vary with the number of pieces of wooden planking. What error would be expected with 100 pieces, end to end? Is it less likely that the errors would cancel out?

A2 What does 'measured to the nearest gram' mean? How accurate is such a measurement? One gram is approximately 0.035 27 ounces. How significant are these figures in the context of the problem?

When does a measured weight of $2\frac{1}{2}$ kg correspond to a measured weight of 2500 gm?

A3 Is it necessary to calculate 1% of 1 foot, or 1% of the length of a standard tennis court? Is it necessary to convert any measurement at all into metres?

Absolute measurements of errors and approximations change when the units are changed. In contrast, the units are irrelevant when only proportions are considered — a difficult point for many pupils to grasp.

A4 Compare **A1**. It is far easier to state the maximum and minimum possible errors, than to state what the most likely total error will be: a perfect example of the simplicity that often appears when questions are asked about extreme cases.

Does the minimum error depend on the dimensions of the original planks? Does it depend on the number of planks cut, under the same conditions? Is the total error more or less likely to be small, if many more planks than three are laid end to end?

A5 Such problems highlight the ambiguity of 'accurate'. In everyday speech it can mean either 'exact, without error', or it can be qualified as 'very accurate', 'fairly accurate', 'accurate enough',....

Does the second sentence ask for the maximum error in each piece, independently? What is the relationship between the errors in the three pieces?

◆

B1 Would the answer to the problem be the same if many more numbers were being summed, but still only one of them was keyed in 10 too low? What difference does it make that the error was in a number being added?

Suppose that each number was keyed in incorrectly. How could the error in the calculator answer be calculated from the individual errors?

B2 In multiplication, the effect of an error depends on the other numbers in the product. Is it possible to calculate the effect of the error without calculating a complete product? Given only that the number '?' in:

$$24 \times ? \times 57$$

is in error by 7, can the error in the product be predicted without knowing the value of the missing number?

B3 What does the problem mean by 'approximate'? This is an extremely ambiguous word, which depends on the context. The problem asks, 'Are you right or wrong?' When is this language appropriate? Under what circumstances would it be correct or incorrect to ignore the error produced?

How relevant is it that it is the last digit that is keyed incorrectly? How relevant is it that the problem is about square roots, rather than squares?

B4 This is a good example of errors which will certainly go in different directions, and will therefore at least partially compensate for each other. Compare a fraction, say $\frac{6}{10}$, as an approximation to $\frac{7}{11}$. The effects of reducing the numerator and denominator by 1 go in opposite directions, and partially cancel each other out.

Is it possible once again, to calculate the error without actually calculating 74×53 or 73×54?

Is the question 'Are you right or wrong?' any more answerable here than in the previous problem?

B5 Is 34^2 a good approximation to 34.7^2? Would the error be greater or less if the numbers were raised to a 4th or 5th power?

How would the relative errors compare if they are considered not absolutely, as arithmetical differences, but in proportion, for example as percentage errors?

What is the difference between 10% and 10.7% compound interest over a number of years?

———◆———

C1 The problem diagram looks much like the similar diagrams for illustrating products such as $(7+5)(4+11)$, with the difference that here the original sides are very large compared to the small errors. This has the advantage of focussing attention on the difference between the original product 15×20 and the erroneous product $15\frac{1}{2} \times 20\frac{1}{2}$, and on the difference between the small but significant strips along two sides and the tiny error that arises from $\frac{1}{2} \times \frac{1}{2}$.

Similar diagrams, in which a large rectangle acquires narrow borders, can be used to illustrate the rate of growth of, for example, the expression x^2, which is approximately (slightly larger than) $2x$.

How does the total error depend on 15 and 20, the sides of the original rectangle? How does the additional small term depend on 15, 20 and $\frac{1}{2}$?

C2 The introduction of P and Q hints at a greater use of algebraic notation, albeit in a simple and relevant context. Pupils who are not ready for such abstraction can still tackle this problem and the next by way of specific examples, and spotting the pattern in their results.

If we assume that, on average, rounding down reduced each number by $\frac{1}{2}$, what would we expect the error in the product to be?

If three numbers were rounded down, before multiplication, what would the maximum error be? There is a comparison here with polynomial sequences, in which the first differences are one degree lower than the original sequence; here, the error is linear when the original expression is of degree two, and so on.

C3 Once again, common sense might suggest that rounding one number up and the other down will give a more accurate approximation. Is this always true? When is this not the case?

C4 Which is greater: the difference in circumference or the difference in area? When is the difference in circumference relatively greatest?

How does the difference in area depend on the actual circumferences of the circles?

If an irregular shape expands by the same amount along its perimeter, how does the change in area depend on the length of the perimeter?

C5 Previous problems might suggest that a box which doubles in volume when the edges are increased by 1 only, must be rather large. Are not the differences in area or volume greatest when the area or volume is large?

Indeed they are, but this problem is about proportion, not absolute difference. Experiment with possible trial shapes will suggest that the proportional increase is greatest when the original box is small.

(The increase in volume of larger and larger boxes when their edges increase by 1 becomes closer and closer to the surface area of the box, which is small in proportion to the volume.)

How many solutions are there in integers? How many non-integral solutions?

What shapes of boxes would increase by a factor of 3 in volume, when their edges increase by 1? By a factor larger than 3?

———◆———

D1 How does the drop in the height of the triangular frame depend on the original length of the beams, if the shortfall is always 4 cm?

How long must the original beams be, for an error of 4 cm to keep them apart, so that there is no triangular frame?

D2 Snooker is a convenient setting for this problem, but it could be set in the context of any ball game. How accurate does Lendl have to be in direction to make a passing shot go within 6 inches of the side lines?

D3 How will the height of a 25-storey skyscraper depend on its age, and the purpose for which it was built?

Compare the error in hanging a door. If one hinge is inset, say, 3 mm too far, at what angle will the door hang? How much smaller will be the gap between the door and its frame? Will the difference be obvious at a glance?

How accurately can such a problem be solved by drawing lines 2.3 degrees apart?

D4 How relevant is it that the plane is flying across the curved surface of a sphere? How far would it have to fly for its initial error to be corrected by the curvature of the earth?

16 Length, area and volume

Pupils have strong intuitive ideas of length, area and volume. However, their intuitions are based on their everyday experience in which they enounter many 'ordinary' objects and very few unusual objects. Alternatively, we might say that when they do enounter something as mathematically extraordinary as a piece of string, they think of it as having length but do not think of it as having either surface area or volume.

Also based on everyday experience there is a natural tendency to suppose that increases in one feature go more or less with increases in another. Once again a sheet of paper, for example, which is of common proportions will increase substantially in area as its length and width increase. The fact that it has a volume will probably be ignored, and the exceptional piece of paper in the form of a long thin strip will not be thought of as 'a piece of paper'.

We also take for granted that an area is the same area, however it is measured, in whatever units. You cannot make an area 'bigger' by measuring it in some cunningly unusual way. Why not?

Mathematicians are interested, as always, in the exceptional as well as the more common cases. They also do not take for granted such intuitively strong and simple ideas as measuring the distance round a curve, or the area of a curved surface, though here the commonsense idea of breaking the object into bits and measuring each bit separately is the basis for a sound theory, and they are aware of the difference between conventions such as the choice of size of unit and relationships between different possible unit which are more than conventional.

These problems probe some of these everyday intuitions and assumptions. It is important that pupils have at hand suitable materials and actual objects with which to experiment.

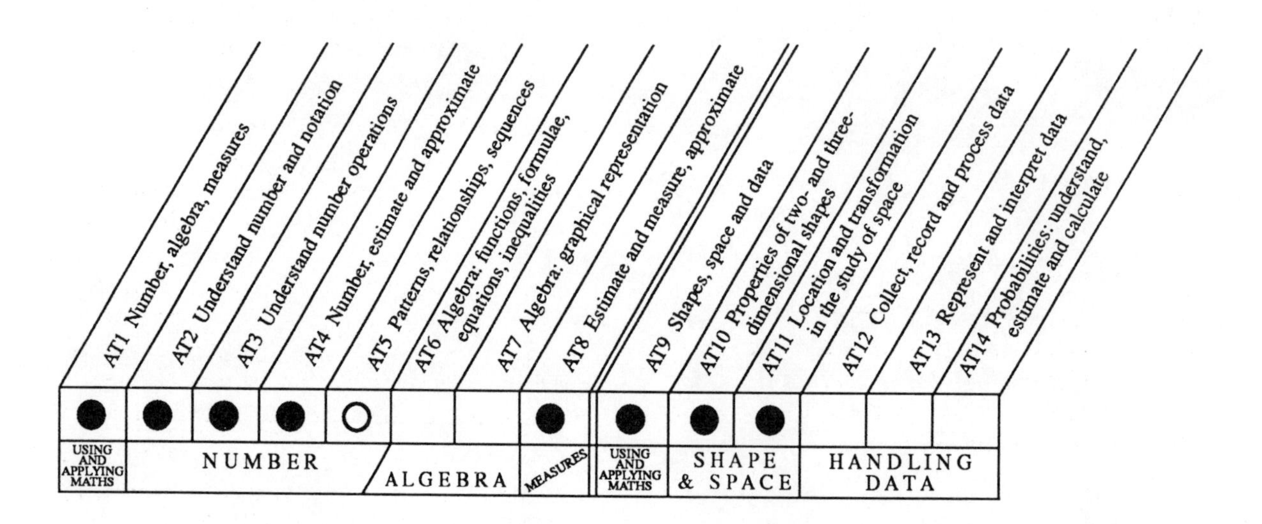

16 Length, area and volume

Comments and suggestions

A1 Such problems immediately raise questions of accuracy. How accurate are the marks on an ordinary ruler? How accurately are inches and centimetres themselves defined? When is '1 inch = $2\frac{1}{2}$ centimetres' good enough? How much more accurate would a comparison be if it were made between the two edges of a longer ruler?

A2 It seems intuitively obvious that 'ordinary' curved lines have a length, just as straight lines do. So the problem amounts to 'What do you mean by the length of a curved line? By what practical means could the length of a curved line be measured? How long is an inch when it is bent round?'

Pupils will almost certainly have come across the circumference of a circle. How is this measured, in practice? How can the perimeters of other curved objects be measured?

A3 Pupils' earliest experiences of measuring curved lines are usually by wrapping thread round cans, which avoids most of the difficulties because the thread can be unwrapped and stretched to a straight line (probably lengthening in the process).

Is it possible to measure a curved line at all with a straight ruler? How accurate are different approximations, bearing in mind that dividing the line into too many small 'straight' bits, introduces its own errors?

A4 The striking feature of this problem is that the line is made up of straight line segments, each of which seems to be one half of the previous segment in length. If this is so, and the the first arm is indeed 6 cm, then the length of every segment is known exactly. Does the fact that they go on for ever mean that they cannot be added up?

How many segments need to be summed to find the total length to, say, the nearest millimetre?

◆

B1 A good practical answer focusses on the shapes of everyday objects whose areas need to be measured. Floors, walls, sheets of paper and so on are more often than not rectangular. Right angles happen to be of vital signficance in pure mathematics also, witness the central role of Pythagoras's theorem. Which of the given shapes will tesselate a rectangle, or will tesselate some rectangles?

Practical shapes tend to have smooth rather than jagged edges. Which of the given shapes when fitted together usually or always produce shapes with uneven edges?

Practical shapes are unlikely to be covered exactly by units of a standard size. How easily can these possible unit shapes be dissected into simple smaller shapes which are a convenient fraction of the whole?

Another approach is to ask what kind of formulas or rules you would need to find areas using different shapes? Which shapes can be generalised to solid shapes which will tesselate in three dimensions?

B2 How easy is it to be accurate when counting in hexagons and parts of hexagons? Is it easier to estimate halves, thirds and quarters of regular hexagons or squares?

What does the phrase 'equally accurate' mean? Surely it is automatically more accurate to use smaller hexagons?

B3 Choosing a unit after knowing what you have to measure is very different from choosing a 'best' unit on general experience. It is common in science and mathematics to adapt the choice to the task in hand.

The problem might be more realistic, easier to appreciate, and the simplest solution easier to see, if the problem presented a large number of similar triangles and asked for the unit which would make it easiest to compare their areas.

B4 This situation is also very different from the problem of finding a 'best' general purpose unit.

What would the answer be if other shapes were added, a circle, and a small unit circle, an irregular shape, and a small but similar irregular shape?

What is the formula for the number of unit cubes in a cube, or unit spheres in a sphere?

◆

C1 What are the two-dimensional analogues of squares, rectangles, equilateral triangles, general triangles, hexagons...?

Which polyhedra will tesselate?

It is easy to dissect a plane shape by drawing a line, straight or curved, with a pencil, and it is also easy to draw a sequence of line segments. How easy is it, in practice, to cut up a solid shape?

There are many striking contrasts between the properties of space of two and three (and four or more) dimensions. The difference in the number of regular and semi-regular plane and space tessellations is one example.

C2 This is analogous to measuring area by dissecting the unit square into smaller and smaller squares.

How easy is it to estimate the volume of a bit of a cube? A good practical challenge is to cut up some cubes, into irregular pieces, and then try to estimate the volume of each piece, either as a proportion of the whole cube, or in relation to each other. Not easy!

Still on a practical note, how could the relative volumes be found easily, if the pieces are made of uniform material?

In theory, it is always possible to measure a volume as accurately as you desire, by using smaller and smaller cubes?

C3 This is easily answered for a cube. What about prisms or other possible units?

How do the volumes of similar shapes in general compare, for example, pairs of spheres which are inevitably similar to each other?

This problem emphasis the rapid growth of volume when the linear measurements are increased. How fast will the surface area of a shape increase, under the same conditions?

——— ◆ ———

D1 How close is the correlation in everyday life between perimeters, areas and volumes? Is it true that the objects whose areas and perimeters we are likely to measure – this qualification excludes pieces of string – tend to be relatively compact?

How might the supposed 'compactness' of an object be measured? See Unit 17, problem **E** (pp. 122–3).

How can the perimeter of any shape be increased without changing its area?

How long are the boundaries of typical countries on a map, compared to their areas?

D2 Length, area and volume are less dependent mathematically than in everyday life. However, the relationship between them is unsymmetrical. The perimeter of a region can be indefinitely large for a given area, but it cannot be too small.

What types of shapes have relatively small perimeters and relatively large areas? Or small surfaces and relatively large volumes?

How symmetrical are they?

How can the perimeter of an irregular shape be reduced, without reducing the area of the shape?

D3 The figures given in these problems are specific, such as 30 cm^3, because this makes the problems rather easier to grasp for many pupils. Other pupils might rephrase them and discuss them in more general terms.

There is a simple analogy with **D2**. What about the solid shape with smallest surface area for a given volume?

D4 Could a cuboid of this volume be any shape we choose? In other words, could it be similar to any given cuboid?

When is a rectangular box most compact?

How many rectangular boxes are there which have the same volume and the same surface area?

What shapes could the box be if all its edges are integral?

D5 How many facts must be known about a rectangular box to fix its shape completely? Is the information given sufficient? Or is there more than one possible solution?

Searching for simple integral solutions is a sensible start.

17 Mensuration, units and scales

Inventing or designing your own scales for measuring something as intuitively clear as steepness or direction is a challenge to pupils which like many of the best problems is easy to appreciate but trickier than it seems. It also involves practical ingenuity and design skills, and links mathematics to Technical and CDT Departments who should also be able to provide the materials necessary and give practical advice.

From a mathematical point of view, these problems force pupils to think carefully about ideas which they may otherwise take for granted. What mathematical concepts can be 'measured'? Is it obvious that can measure 'steepness' at all? If you can measure steepness, can you also measure curvedness, or roundness?

They also highlight the difference between conventions, such as the length of a foot or a metre, which have social and historical origins, and mathematical concepts, such as the idea that if a scale is uniform — what does that mean? — then it is defined by two points on it, such as the zero and unit marks.

There is also an obvious relationship with elementary science. It is helpful if some scientific instruments can be borrowed and studied, in actual use, to see how they work and how their units and scales are designed.

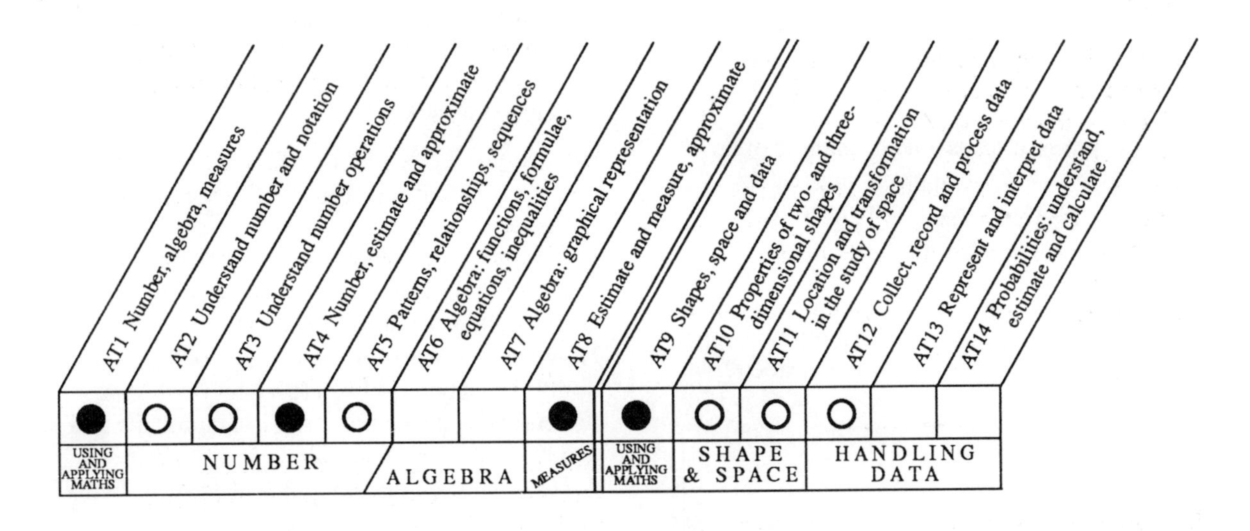

17 Mensuration, units and scales

Comments and suggestions

A The emphasis on short periods of time in this problem and on small objects in problem **C** is deliberate, to emphasise accuracy and the difficulty of avoiding errors.

The question naturally arises, how long and heavy should a pendulum be if it is to have a period of exactly 1 second? Or 2 seconds, or $\frac{1}{2}$ second? What is the relationship between the weight and shape of the pendulum and the length of its string, and its period? These questions are scientific, but mathematics is needed to describe and analyse their answers.

What other devices could be used to measure time? Could clockwork or electric motors be used?

◆

B 'Steepness' is an excellent example of an everyday concept which pupils appreciate intuitively, and which can be measured mathematically in a way that fits pupils intuitions.

It is easy enough to agree that the first ladder is steeper than the second ladder, but how much steeper is it? A bit steeper? Twice as steep?

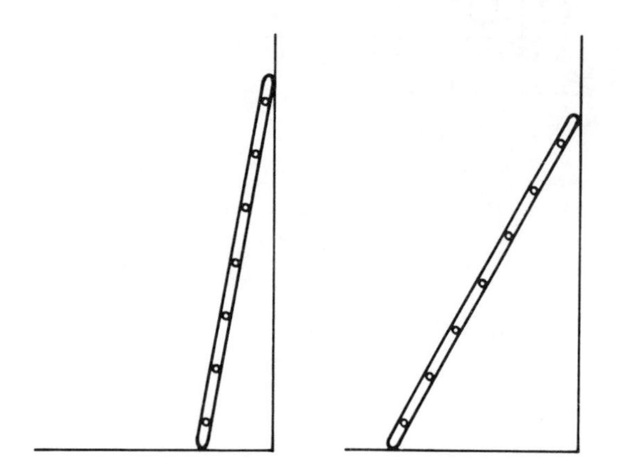

If pupils suggest different ways of measuring steepness, then the questions naturally arise: what is the relationship between them? How can one measure be turned into another?

How is the steepness of a hill measured by conventional road signs?

◆

C This is also a scientific problem which soon involves mathematical concepts. From what kind of materials should a delicate weighing machine be made? What physical principles, or everyday observations suggest how a delicate weighing device might be constructed?

It is easy to make a device which will respond when, for example, a very small piece of paper is placed on it, but how can numbers be associated with this response? What will the units be when the object is too small to move the kind of weighing device that pupils are accustomed to?

How can a scale be designed which will tell you when one object, for example, weighs twice as much as another?

◆

D The radius of an ordinary protractor is too small to allow accurate measurement of angles. How much more accurate would a protractor of double the diameter be? How accurately can directions be measured using a sighting line and a much larger circular scale?

How is it possible to mark a scale accurately in the first place?

A surveyor's theodolite is a practical solution to this problem which is also portable. How does it achieve the advantages of a large scale in such a small instrument?

◆

E This is a good discussion problem, not least because it is ambiguous. Does it mean 'smoothly rounded' like an ellipse, lacking sharp corners, or does it mean compact, in which case a square scores highly? Does it refer to a local property of the shape, its smoothness as the eye follows it, or to an overall, global property, such as being convex, or looking more or less like an agreed round shape such as a circle?

Is it possible to construct a measure of 'roundness'? How would, or should, shapes such as these score, which combine an overall kind of roundness with a multitude of sharp edges?

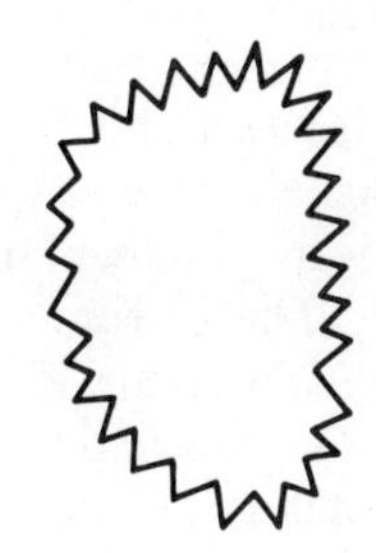

What would or should a round shape in three dimensions look like?

F Is it true for a letter balance, like this

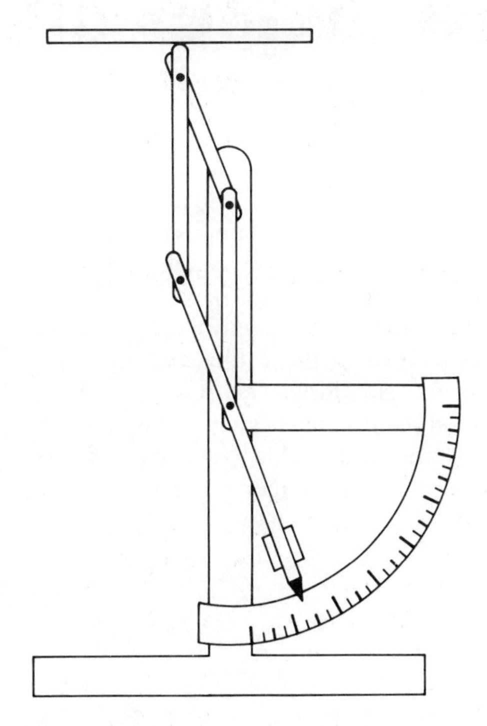

Is it true for scientific instruments such as voltmeters and ammeters? Is it approximately true for many instruments?

If a scale is not uniform, is it possible to predict all the other points on it if you know, for example, the positions of 0, 1 and also 10?

18 Scales and balances

These problems about balance and centres of gravity involve ideas of ratio and proportion in a practical context which pupils can explore and which can easily be provided in the normal mathematics classroom.

They also relate directly to an aspect of science in which the mathematics involved is both very simple and very powerful. Science laboratories should have sets of weights and workshops will be able to provide suitable pieces of wood or other materials.

Many of the problems are expressed in quite specific and simple terms. Pupils should appreciate that these questions are only representative of the same problems with different and more complicated weights. As usual, there is big increase in difficulty as soon as the numbers become more complicated.

These problems deliberately do not involve interpreting weights on a balance in terms of an equation, and the balance is not used as an analogue device for solving equations. This is not one of the themes of this section.

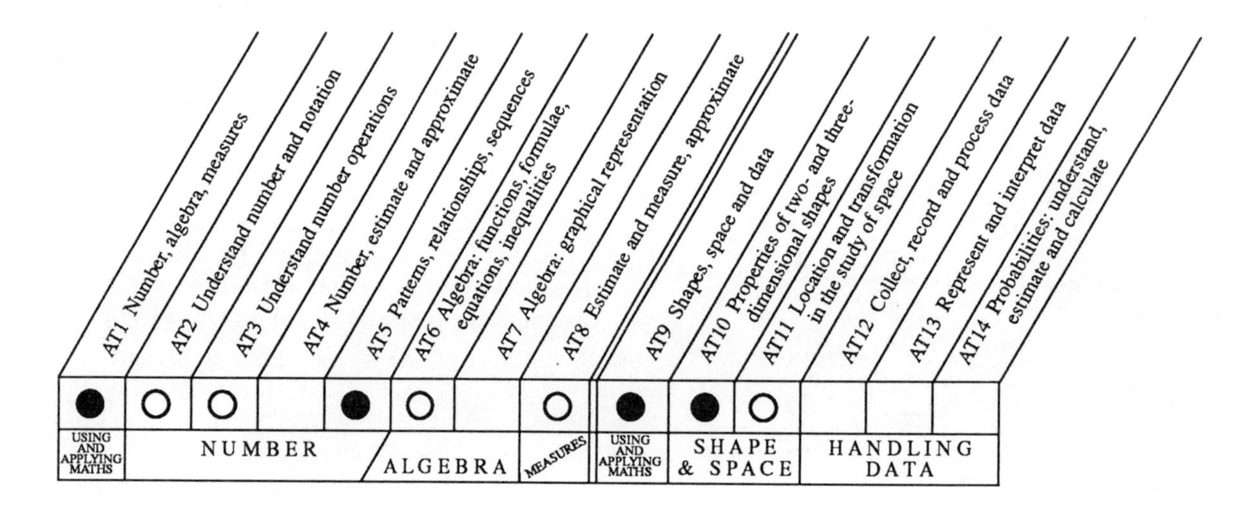

18 Scales and balances

Comments and suggestions

A1 When small integral weights are used, the balance point is at a more 'obvious' point on the rod. Without such a simple start, and bearing in mind the weight of the rod and some inevitable experimental error, the balance point may not appear to be any special or identificable point at all.

It may help if the length of the rod, (actually the distance between the hooks on which the weights are hung,) is a suitable round number such as 60 or 100 cm.

Although 'rod' suggests a circular cross-section, a strip whose vertical cross-section is a rectangle, narrower than it is tall, will have greater resistance to bending, for the same area of cross-section.

Where would the balance point be if the weights were 1 kg at one end and 3 kg, or 4 kg or 5 kg... at the other end? How does the balance point 'move' as each weight in this sequence is replaced by the next weight?

A2 Successful solution of **A1** may suggest the kind of result to look for, while leaving some doubt as to how accurately the experiment will fit such a very simple mathematical pattern. It helps that the weights specified here are greater than in **A1**, so the weight of the rod is likely to have a negligible effect.

How obvious is it that the balance point will be nearer the heavier weight?

A3 A metre of so of the thickest wooden strip or dowel rod can be cut to weigh exactly 1 kg.

This problem is far harder than problems **A1–2**. How does the rod balance by itself? How does it behave when it is used as a weight and hung from another rod, for example against a 1 kg weight?

A4 As usual, more complicated numbers raise difficulties, partly because they are less familiar to pupils and therefore harder to think about, and partly because the arguments and proofs that work for simple numbers are immediately much more complicated.

Intuition, and some general concept of continuity may suggest that the balance point 'ought to be' at the point which divides the rod in the ratio 2.34:5.6.

Will experiment confirm this? How accurately?

Does it help to think of these weights as respectively 234 and 560 units?

A5 This problem offers another way of thinking about weights on a rod. Instead of thinking of the weight at the ends and the balance point between them, it presents the weights as placed at certain distances from one specified end and asks for the balance point.

This might suggest starting with the weights near the given end and experimenting to see what happens to the balance point when they are moved further away from it.

How much do weights in particular positions 'pull' the rod down? How could this pull be measured?

◆

B1 This is another way of looking at the same situations. Instead of giving the weights and asking for the balance point, the balance point is given and the relationship between the weights and distances is demanded.

How do the distances of the weights from the balance point compare?

B2 Simply comparing the ratios of the distances and weights is no longer sufficient. In fact, only one of the ways of describing the solution to problem **B1** can easily be extended to explain this balanced arrangement.

This is a typical example of how further investigation forces the mathematician to focus on one way of looking at the problem, and highlights concepts which turn out to be important elsewhere.

What other combinations of three or more weights will balance?

◆

C1 This is a good problem for discussion, because pupils have difficulty in intuitively predicting the answer. A piece of string can be looped over the strip, 10 cm from the chair back, so that the distance of 10 cm is clearly seen, through it may be difficult to decide exactly where the strip is balancing or rocking on the chair back. It is quite possible that the string will break if it is not strong enough.

How does the strength of the pull down, needed to lift the 2 kg weight, increase as the point of pull is moved nearer to the back of the chair?

Is it possible to lift the weight by pulling down only 1 cm from the fulcrum?

How easy is it to lift the weight if the situation is reversed, and the weight is 10 cm from the chair back and you pull down at the far end?

C2 How does the ease of lifting the weight vary if it is moved towards or away from the end which is being lifted?

What would a graph of the 'apparent weight' of the 2 kg look like, as it moves from the fixed end to the movable end of the strip?

What difference does the weight of the wooden strip make?

Spring balances are useful for such experiments.

——— ◆ ———

D1 A triangle can be cut from plywood. Will it make a significant difference if it flexes a little?

In this problem and the next two, the term 'balance point' is replaced by the more standard term 'centre of gravity'. Note, however, that the latter is in many ways a more abstract term. 'Balance point', at least as used above, refers to a point found by actual physical experiment, in which the fact that the arrangement balances can be literally felt as well as seen, as well as a point whose position can be predicted by calculation. These actual experiments are vital to develop pupils' intuitive feeling for balance.

'Centre of gravity' has far less concrete and physical feeling.

How can this problem be solved by considering the weights two at a time?

D2 It is easy to experiment with different shapes of triangle cut from card. Is there any connection with the previous problem?

How can the centres of gravity of other polygons be found?

D3 The L-shape can be seen as made up from two simpler shapes – for example, two rectangles. Where are their centres of gravity? What is the connection between the centres of gravity of two shapes, or objects, and their centre of gravity when they are rigidly fixed together?

What is the centre of gravity of this shape by experiment?

Can the centre of gravity of an object lie outside the object?

D4 How will regular shapes hang from a thread? How will a triangle hang, bearing in mind the solution to **D2**?

How many tests of this kind are needed to find the centre of gravity of a plane shape?

What about solid shapes?

19 Percentages

Most of these problems focus on the more curious and surprising properties of percentages. They are not intended to practice techniques that pupils have already learnt. These are best exercised on practical investigations, such as **D1** and **D2** hint at, which cannot easily be presented in the pages of a book or on a worksheet without a great loss of motivation and meaning.

Problems **A1** and **A2** refer to hundredths rather than percentages for a simple reason. Pupils who are not familiar with handling hundredths gain nothing and are only likely to be confused by introducing the term 'percentage' prematurely. As usual, the new language should match the pupil's experience, not rush ahead of it.

There are no problems on practical investment, bank loans, hire purchase, and so on, because such problems are best presented with realistic figures and realistic situations, preferably with the actual documents that would be used. There is no comparison in terms of motivation and understanding between the vividness of actual documents or excellent imitations, and crude imitations or 'simplified' versions which lack all realism.

The problems from **B1** onwards have been aimed at the pupil's intuitive feeling for percentages, because this is so often weak, and because some features of percentages are strongly counter-intuitive. They are especially concerned with change, increase and decrease, and an active and dynamic view of percentages.

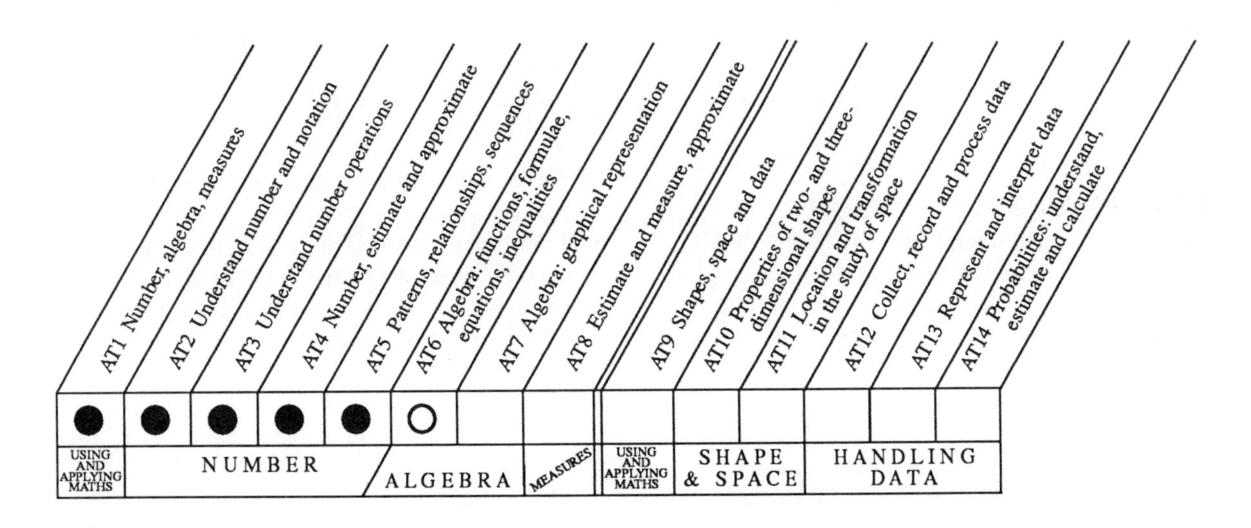

19 Percentages

Comments and suggestions

A1 This problem is essentially about the properties of numbers. What fractions can be expressed exactly as whole numbers of 12ths? Or 60ths? Or 23rds?

The language 'thirds of hundredths' is complicated, but only illustrates the conceptual difficulties which pupils face, 'thirds of one percent' does not get rid of the difficulty; it just transforms it.

Which fractions can be expressed as thirds, plus hundredths of a third?

A3 This is an interesting problem in pure mathematics, which also has practical use and involves pupils' intuitive feeling for fractions.

What could or should 'simpler' mean in this context? Is a fraction with a smaller denominator necessarily simpler than one with a larger denominator?

How 'simple' can you expect a fraction approximately equal to 87% to be?

How simple can a fraction be that is an approximation to $\sqrt{2}$, or π?

A4 A first step for most pupils will be to think of particular fractions which fit the condition of the problem.

How can a fraction be approximately equated to a percentage?

Dividing the denominator into the numerator is the last method that will occur to pupils who have not been taught a standard technique. What intermediate steps might lead them to appreciate why that, typically powerful and universal, method works?

◆

B1 This poses serious difficulties of language for pupils before it poses a mathematical problem. So do **B2** and **B3**. Such difficulties can be explicitly presented as part of the problem. The first step, as so often, is to discuss and decide what the problem is talking about.

Why is the increase in Miss Watson's money not equal to the sum of 10% and 12%? Why is it 'obviously' slightly greater?

B2 This problem, and the next, make excellent discussion material, because they are counter-intuitive for so many pupils. Surely Mr Wilson should be back where he started after the two years?

What percentage loss would indeed leave him back where he started? What is the relationship with reciprocals?

B3 The same pupils who think that Mr Wilson in **B2** should end up where he started will often think that one of Alison or Jack must be better off. Compare Unit 11 problem **D1** (p. 84).

Why are the situations different?

Is it true that in any sequence of percentage increases, the order of the increases is irrelevant?

◆

C1 In scientific applications and for any financial purposes, it is the geometrical increase when the interest is not removed which is important.

Once again, the results are often surprising, especially if pupils estimate (or guess) the answer first.

What difference would it make if the starting sum had been £10, or £40?

Why does the rate of increase of the total over the years increase also, when measured as a difference?

How quickly can the solution be calculated?

C2 'Double Your Money' is an easily appreciated idea. The doubling time for a given rate of increase also makes rough estimation much easier, because if you know the doubling time, you can easily work out the time taken to multiply the original sum by 4, 8, 16,..., 1024 (approximately 1000)... .

In problem **C1**, the doubling time for an increase rate of 5% is between 14 and 15 years. Is there any simple connection between this figure and the doubling time for a rate of 10%?

C3 This problem expresses in specific figures the problem of interest compounded at smaller and smaller intervals. Does making the intervals smaller and smaller and smaller... increase the final sum indefinitely?

D1 Historical statistics often mean little to pupils. The cost of living, however, has changed greatly during their lifetime, and is also a topic of interest to their parents and other adults who always have recollections, usually distorted, of the good old days when everything was so much cheaper! But what were wages in those days? If everything was so cheap why did not everyone live like millionaires?

How accurate are the prices quoted for 1968? Were they constant throughout the year?

Why do some objects increase in price faster than others? Why do prices increase at all? Why do prices sometimes fall?

By collecting prices from informants, or from published records, and then drawing a graph of changes in typical prices, what would the expected price of a bar of chocolate, or a new family car be in the year 2000?

D2 There is no substitute for producing the actual carton in class. Such special offers are easy enough to collect and investigate. What are they really worth? Which of the competing offers is best?

Similar problems arise in comparing the price in a supermarket of a large Gleamo washing powder, the even larger family packet of Gleamo and the giant economy Gleamo. How much is saved by buying the giant economy size?

20 Diagrams, charts and graphs

Problems emphasising the meanings of charts and graphs, their great variety and the difficulty of designing them so that they are as clear and informative as possible are well suited to group work and discussion.

Motivation is increased when the data are relevant to the pupils. Good sources are pupils' own interests, data from the pupils' own scientific experiments or projects in other subjects, and events in the news – the explosion at Chernobyl as I write, following the coldest winter for years and the preceding World Cup.

Motivation can also be enhanced by presentation. Milk, the plausible solution to problem **A3**, becomes immediately more relevant if there is milk in the classroom, being drunk perhaps by the teacher in a cup of tea. Biscuits increase the relevance still further. What have biscuits and milk in common? Fat, carbohydrate, protein and water. How much?

These charts and graphs are relatively complex. This is deliberate, since the object is not to answer questions based on previously learnt facts, but to approach the graphs and data as problems to be thought about and puzzled over. Far from being atypical, this complexity is very common in real life. The simplest bar charts and graphs are rare and the ability to read them is insufficient to interpret many charts and graphs in newspapers and magazines.

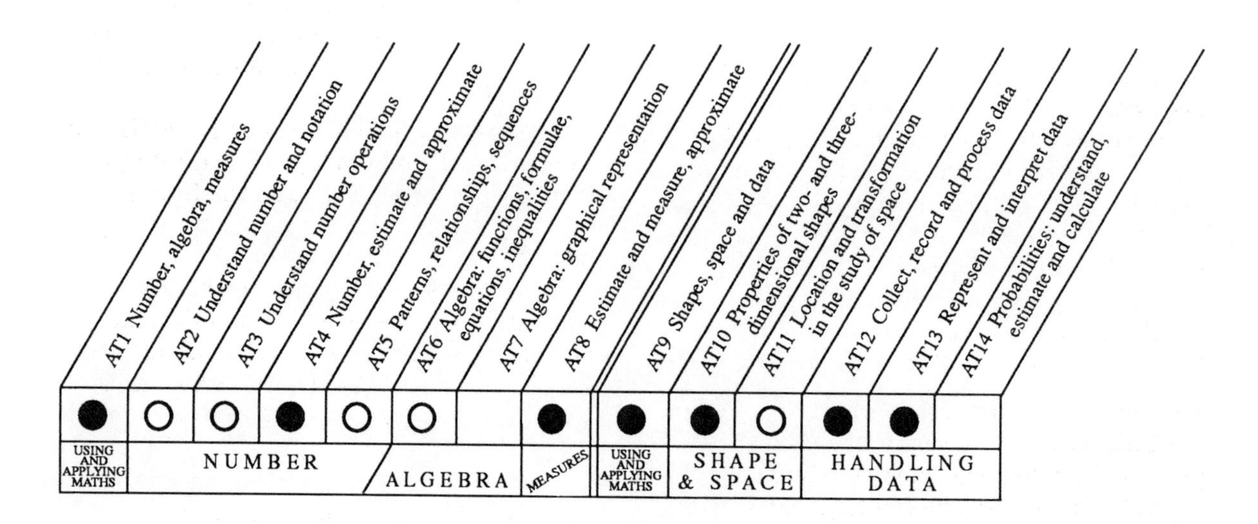

20 Diagrams, charts and graphs

Comments and suggestions

A1 The graph shows a periodic phenomenon. What is the commonest period of such phenomena? How do pupils change during a period of, say, one day? What periodic phenomena are familiar to pupils?

What does the shape of each portion of the graph suggest?

A2 This drink contains substantial amounts of three different substances, and small quantities of a fourth. Is the greater part of the drink water? What are the compositions of popular soft drinks? What are the compositions of tea and coffee?

How could a drink be designed by a manufacturer to have this kind of contents?

A3 Pupils can work out the meaning of a Venn diagram, like most charts and graphs, by commonsense and discussion. Having worked out the meaning for themselves, they are more likely to remember and understand it.

What other information could be put into a similar diagram? How else might a diagram displaying the same information be drawn?

How could the same or similar information about members of the class be displayed?

A4 This problem is much more difficult, but it illustrates how the same information can be presented in more than one way. The choice of presentation may be important in making conclusions easier or harder to draw. In the second presentation here any group of foods with similar composition will leap to the eye as a cluster, while a single food with very unusual composition will be equally identifiable.

A5 This nomogram is taken from *Maths at Work* by Geoffrey Howson and Ron MacLone (Heinemann Educational Books, 1983).

How accurate is a nomogram? Which other adults with different combinations of weight and height would have the same Body surface area according to this chart?

How accurate does a nomogram have to be? Why and when are nomograms useful in practice?

Is it possible to construct a nomogram for finding the surface area of, say, a rectangle, from the lengths of its sides?

Could a nomogram be designed to show the average of two numbers?

A6 Like many charts and graphs in published statistics, even adults who only occasionally consult such presentations may have to study the chart in order to work out what it means and how it should be read.

Does this chart present its information clearly? How else might the same information be represented?

What trends in the length of paid annual holidays does the chart suggest?

A7 An excellent example of a lot of information presented very clearly, (only assuming that the first column represents accidents between 12.00 and 1.00 a.m.).

Why is the variation for Saturday and Sunday smoother than for Monday to Friday? Why is there variation at all in the accident figures? Why do the peak times for accidents vary between weekends and weekdays?

Does the probability that if you have an accident you will be killed or seriously injured vary through the day? Are you more likely to be killed or seriously injured on a weekday or at weekends?

◆

B1 Such a problem is a problem in design, in graphic design, as much as mathematics. What features of the information do you want to show? How can they best be brought out by the design?

Which is preferable: a bar chart in which each child is represented by two bars in different colours for height and age or a scatter diagram? What other design might be superior to either of these?

What trend would be expected in this information? Would there be a striking difference between boys and girls? How could the presentation be designed to emphasise trends of sex differences?

Would a design which showed the variation in height as clearly as possible, be able to show variations in age as clearly?

Does the 'final' presentation show features which were unexpected, and which might be shown better by a different presentation?

B2 Why do the wind speeds not increase more smoothly with the Beaufort numbers? How are wind speeds measured in practice? Do most people who refer to wind speeds actually measure them with an instrument?

B3 Two-dimensional tables can be simplified when there are patterns in the data. For example, lines could be drawn through equal values. Why is there only one maximal value?

How 'approximately' should the data be displayed? For what purpose? There is a trade-off here, as so often, between information and clarity.

B4 Sport is a fruitful source of statistics, as well as illustrating the difficulties of realistic extrapolation and the extraordinary capacities and limitations of the human body. The present mile record is equivalent to running 17.6 hundred yard races, consecutively — in how many seconds each? How nearly can a middle distance runner turn a mile race into a continuous sprint?

If the data only went up to, say, 1960, how easily could the future drop in the mile record have been predicted? What does this suggest about the reliability of extrapolating to the year 2000?

B5 What significant changes have taken place in the distribution of the population over 130 years? How can these changes be highlighted by the design of the diagram?

What is the present distribution of the population? Is the distribution still changing or has it stabilised?

21 Inventing coordinates

It should be a reasonable assumption that pupils will have surreptitiously come across examples of numbered square grids, for example through playing games, long before they need to have any formal knowledge of them. There are therefore no problems here on the actual 'definition' of Cartesian coordinates.

These problems are concerned with developing some feeling for the differences between the surfaces of several important shapes, as much as developing intuitive feeling for coordinates. There is a large difference between looking at a cylinder as a curved object, or at some shape for a container, and actually examining its surface closely enough to discover how it can be covered with a numbered grid. By focussing on the idea of coordinates and the surfaces of shapes, pupils are prompted into thinking about each in new ways.

By way of contrast, the study of other simple surfaces illustrates the extreme simplicity of the Euclidean plane and its usual coordinate system and helps to develop pupils intuitive feelings for its properties. As usual, features that might otherwise be taken for granted, stand out in a comparison with other surfaces which do not share them.

It is essential that pupils have the shapes and surfaces to handle and explore. Any kind of ball will do for a sphere. Large cylinders used for packaging are fine, or pupils can make them for themselves. Cones may have to be constructed, which is an interesting problem in itself.

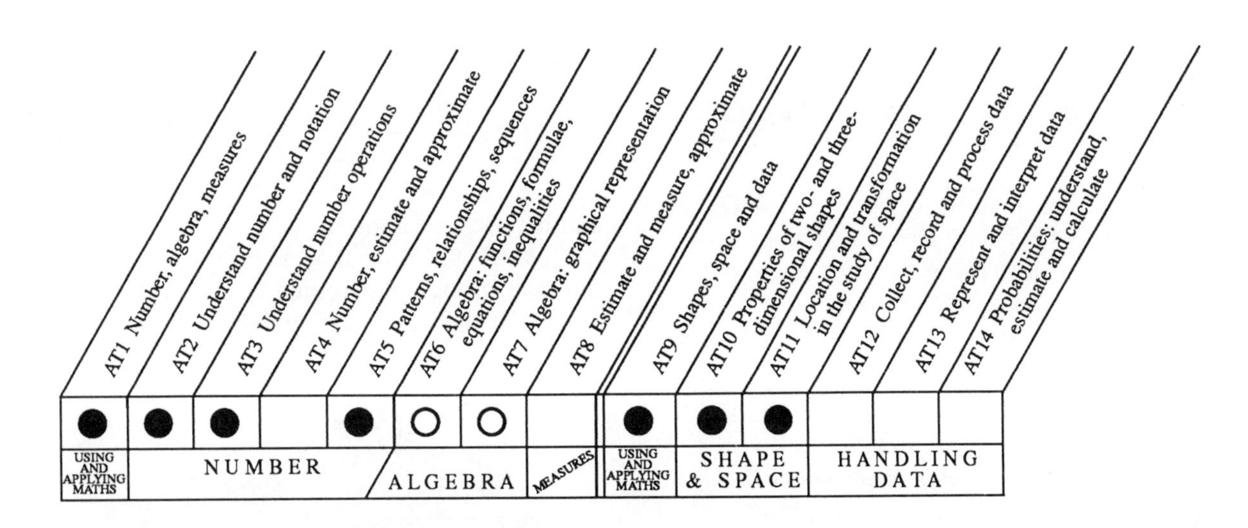

21 Inventing coordinates

Comments and suggestions

A1 What are the essential features of a coordinate system? Is it sufficient that every point should have a label, or is it necessary that no two points should have the same label?

How do different variant coordinate systems compare with the Cartesian system? How many essentially different variants are there? In what sense is the Cartesian system simpler?

Do all coordinate systems for a plane have two sets of intersecting lines? Why two?

A2 Cylinders in everyday life are almost always circular. Mathematically, they can have any cross-section. (Or can they? Can the cross-section of a cylinder intersect itself? Could it be a figure of eight?)

Does it make a difference to the coordinate system if the cross-section is not circular?

The ends of a cylinder are also more likely to be included in an everyday context, than in a more mathematical context. Could they, or should they be incorporated into a coordinate system?

How does the curved surface of a cylinder differ from a flat plane? How could it be constructed from a plane? Can any coordinate system on a plane by converted into a coordinate system for a cylinder?

A3 A cone is less symmetrical than a cylinder. What about the unique and special point, its vertex? Can this be given, should it be given, a special position in the coordinate system?

A4 The system of longitude and latitude has two special points, the poles. This might seem unnecessary, since the sphere itself has no natural special points, unlike a cone. Is it possible to design coordinates for a sphere without any special points?

A plane coordinate system was easily wrapped around a cylinder or cone. Can it be wrapped around a sphere, perhaps with a little distortion only? Or is a sphere in some way essentially different from a plane?

A5 Inner tubes from bicycles or cars are in the form of a torus. So are the rubber rings used by children at the seaside.

A torus is very different from a plane or a sphere. For example, a ring round a torus need not divide the surface into an inside and an outside. In fact there are two such kinds of rings. These rings cannot be contracted to a point, unlike a circle on a sphere or a plane which we can easily imagine getting smaller and smaller.

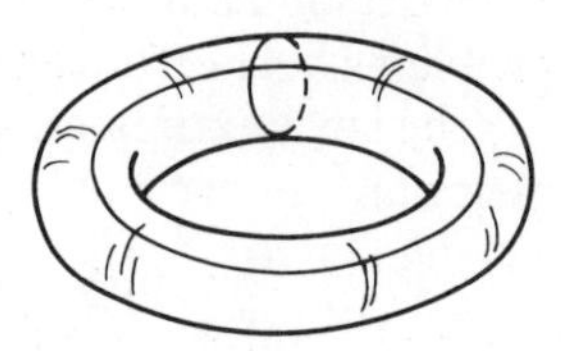

On a cylinder a ring can be drawn round the cylinder which cannot be contracted to a point. How does a cylinder compare to a torus? How could a torus be made from a flexible plane sheet?

How symmetrical can a torus be? How can this symmetry be exploited to design a coordinate system that is 'natural' and simple'? experimental design

---

B1 Coordinates can be used to transfer any shape or set of points in one system on to another. D'Arcy Thompson in his classic book *On Growth and Form* showed how the shapes of different species of fish could be transformed into each other by distorting the coordinate system.

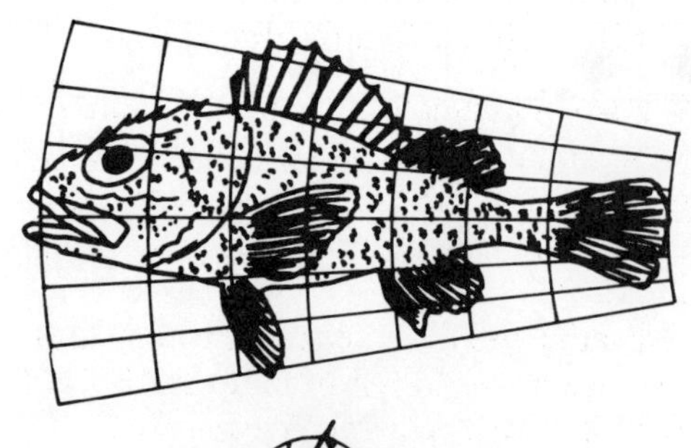

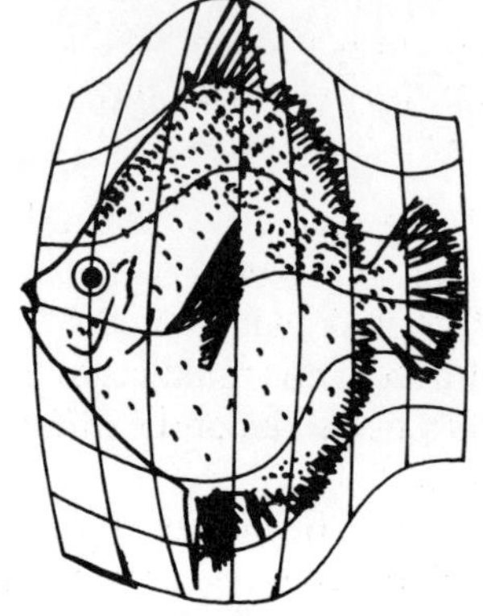

In what ways are simple shapes such as straight lines distorted? A straight line on a plane is the shortest distance between two points. Will the transformed line be the shortest distance between two points on, say, a cylinder?

Will a set of points forming a closed loop necessarily form a closed loop when they are transferred to another surface? Could a straight line on a plane become a circle in another system?

The given set of points was chosen for its simplicity. Is it even possible for all the points in this sequence to be transferred to some other coordinate systems?

B2 It is useful to draw the coordinate system for a plane map on a transparent sheet which can then be slid over the map. (A transparent cylinder will slide over a cylindrical map.) This problem does not state that the two grids were in the same orientation, though that assumption is necessary to bring the problems within range of almost all pupils.

What happens if two systems have the same origin and orientation but the scales on their axes are different? What if the axes and origins are in the same positions, but the order of the axes is switched round?

B3 This is an invitation to think of analogies with coordinate systems in two dimensions. Since three-dimensional space is harder to visualise and think about there is even more incentive to think of simple and straightforward systems.

What is the three-dimensional analogue of an ordinary square grid? How can the more or less familiar longitude and latitude be adapted to describe positions in three dimensions?

B4 Polar coordinates manage to describe the position of any point on a plane without using any negative numbers. Why is it worth using Cartesian coordinates which need negative numbers?

Negative numbers seem to be needed, at most, when a surface is infinite in at least one direction, like a cylinder. Are negative coordinates ever necessary for describing positions on a finite surface, like a torus? Are they ever useful for this purpose?

B5 Games on a square grid depend on simple geometrical or topological properties of the grid, usually including its boundary. Playing games on different coordinate systems is a good way to experience some of the differences between them.

Chess has often been played on cylinders or toruses, by identifying one or two pairs of opposite edges of the board. In this position the white bishop can capture the black knight!

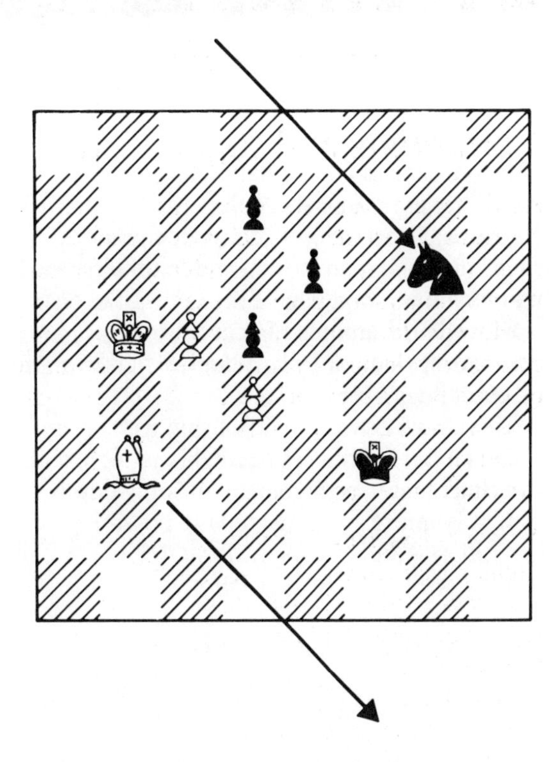

The oriental game of Go can also be played on different grids, including grids that are not square, although the game usually changes markedly because it is so dependent, much more so than chess, on topological features such as the boundary and the corners.

Such games can also often be adapted to three-dimensional play, which also tends to change then drastically, because each cell or intersection tends to be adjacent to so many others.

What will count as a straight line on unusual coordinate systems? Diagonal lines are allowed on the usual square grid. What does a diagonal look like on different coordinate systems?

What difference do special points such as the vertex of a cone make to the game?

How can grids for games played on unusual surfaces be drawn on an ordinary sheet of paper?

B6 Shapes can also be moved across, as it were, a coordinate system, by simply adding numbers to their coordinate. A set of points on a Cartesian system does not change its shape when this happens. It is merely translated. Do sets of points on other coordinate systems keep their shape under this transformation?

If such transformations do not preserve shape in general, do they preserve shape when only a small part of the surface of, say, a cylinder or sphere is considered?

22 How fast does it move?

Many pupils have a poor intuitive understanding of what is happening when an object moves. Practical problems give pupils a chance to observe closely, to make the same experiment again and again, and to understand better what is happening. These problems also provide a large amount of practice in calculation, whether by hand or by calculator, in a challenging context.

There is an obvious relationship with science. The essential equipment needed here for experiments is very simple, but if some pupils want to be more sophisticated the possibilities are endless. Equipment may have to be borrowed from the science laboratory or designed by pupils.

Even at the level demanded here there are genuine simple problems of experimental design — for example, in deciding how to measure the angle moved through by a pendulum.

None of these problems relates directly to acceleration, which is proportionately harder than the concepts of speed and velocity. The opportunities are there for pupils who think that far.

These situations have been chosen not only for their simplicity, but also because they are extremely visible. Equally clean and accessible experiments (such as timing the speed with which a small electric motor can lift a weight, as the current varies) are less visible.

The formulae in **C1** and **C2** are chosen, as usual, to be relatively easy. Other quadratic formulae lead to ideas similar to those encountered in 'An example with commentary: quadratic equations' (pp. 16–21) from a different perspective. Thus, if the coefficient of the squared term is fixed, then all the graphs are the same shape and the speed at which the expression changes is 'essentially' the same, merely shifted with respect to time.

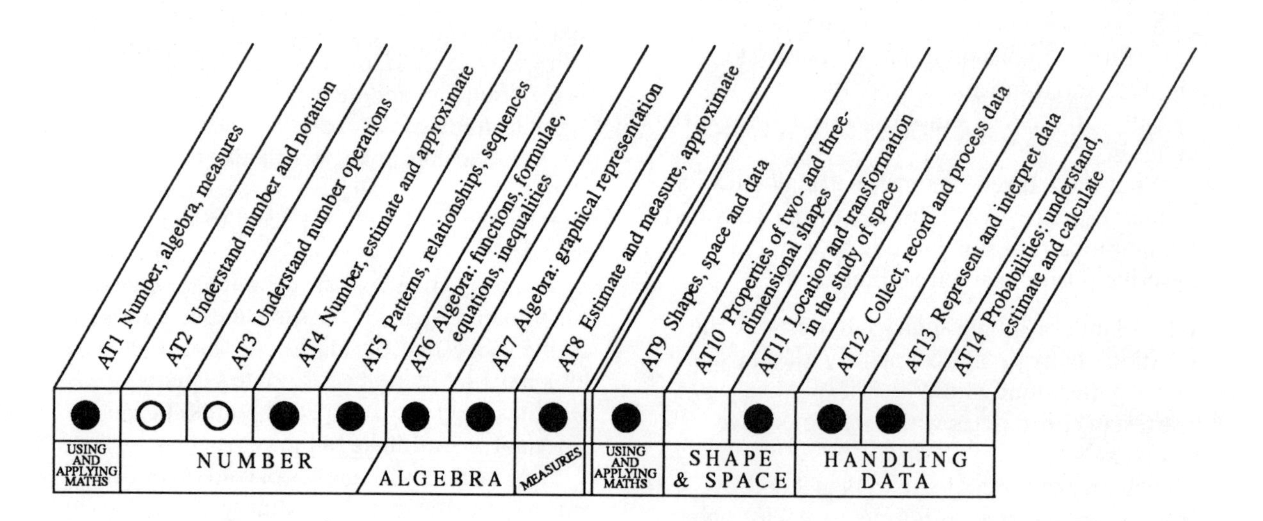

22 How fast does it move?

Comments and suggestions

A1 'Fast' is ambiguous. Does it mean rotational speed or linear speed? What is the relationship between the two?

How fast do different points on the record move, when their speed is measured linearly, and as rotations per minute?

A problem such as this raises the question of the difference between speed and velocity.

A2 Most children have experienced a Big Wheel, including the feeling in their stomachs due to the acceleration of falling and rising. A model is easy to make with an electric motor or the record deck of problem **A1** turned on its side.

The emphasis on height suggests that speed or velocity might be considered in different directions.

How does the variation in height compare with the side to side motion of a pendulum?

A3 A well-known puzzle asks, 'Which parts of a train travelling from London to Brighton are actually travelling from Brighton to London?' This appears paradoxical or impossible as long as it is assumed that it is the same point which is travelling backwards to London, rather than a succession of different points.

Points on the flange of a wheel are temporarily moving backwards for a portion of every revolution. What other parts of the engine might be moving backwards during part of their cycle?

A model can be made from cardboard and one of the cylinders used for despatching posters. How does the path of a point on the flange vary with its distance from the centre of the wheel?

A4 The marble is intended to represent any object, such as a freewheeling cart or a toboggan, which is rolling down a plane. The marble happens to be very reliable and is easily rolled down a length of angled strip.

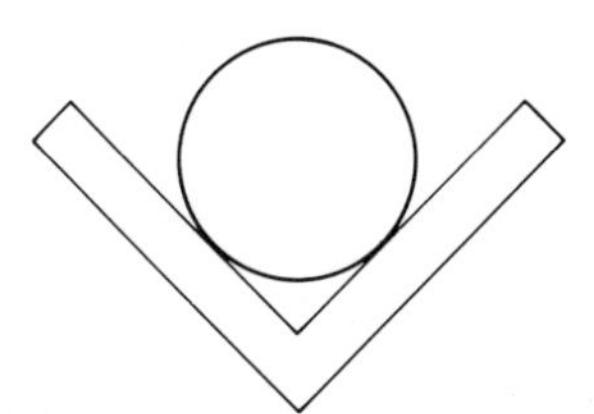

How does the time for one descent vary with the slope of the chute? How does the time vary with the length of the chute?

How accurately will an ordinary watch with stopwatch facility measure the time of descent?

How does the fall of the marble down the chute compare with an object falling vertically through the air?

A5 The wording of the problem is intended to suggest that these two motions are very similar, as indeed they are, a fact which Galileo realised. However, the path of a marble on an inclined plane is much easier to follow with the eye. It may also, with care and attention, be recorded by inking the marble before projecting it across a sheet of paper.

How does the velocity of the ball vary in a vertical direction? Horizontally?

In what respects is the flight of the marble or projectile symmetrical? Why is it not entirely symmetrical?

By how many different paths could, for example, a table-tennis ball be thrown from a fixed point onto a target? Which paths will be shortest, in distance or time taken?

B1 How can speed, in general, be deduced from information about time and distance or a time–distance graph? How much information does a graph like this contain? How much does it suggest, if some simple and reasonable assumptions are made? How many different interpretions of the graph are equally plausible?

How can a graph be constructed from the information in this graph, to show the speed of the cyclist as a function of time?

B2 If it does show the displacement of a pendulum, then for a change the time axis is vertical, emphasising that the usual horizontal time axis is only a convention.

What other phenomena have similar graphs?

C1 What does it mean to ask how fast a sequence is increasing? Pupils naturally tend to look at differences in a sequence. If the first differences are thought of as speed, what do the second differences represent?

Does it make any difference if a sequence such as this is drawn on a graph?

How can the rate of increase be described most clearly in words?

C2 The expression $100 - t^2$ is chosen to relate directly to **A4** and **A5**, though it is not necessary that t should represent time.

How does this sequence and its rate of change compare with that of **C1**?

'Fast' is again ambiguous. It could mean the decrease between two integral values of t, with no conception of an average rate of decrease, or rate of decrease at one point in time, in which case the data in the table could be seen as representative of the values of $100 - t^2$ for every possible value of t.

C3 x^3 has been chosen as a very simple representative of higher-degree polynomials. This problem and **C2** are closely related to Units 7 and 9. However, the emphasis on speed, which varies continuously, and the presence of half-integer values for x, suggests that all values of x might be considered, not just integral values.

Which is most sensible: to calculate a rate of increase by using integer values only? By taking differences also from one half-integral value of the next? Or by taking these values as only representative of all the possible values of x^3?

A small point which is likely to be spotted: the decimal parts of the half-integer values go through a cycle of four values only.

C4 There is no suggestion here that only integer values should be considered, though for large values of x it is enough to consider integer values of x only.

Does the expression 'as x increases' necessarily imply that x is large and getting larger, or could x be small but getting larger?

Which of these functions increases fastest when x is small, say, less than 1?

Which polynomial expressions in general increase fastest in value? In the short term? In the long term? Which increase most slowly?

23 Making sense of data

These problems are designed to force pupils to think about data, what information they can get from data, and what it 'means'. They are intended to focus at least as much on qualitative as quantitative aspects.

The data in these problems should be regarded as samples only of the kind of data that can be used. In statistics there is immense value in using data generated or chosen by the pupils themselves — for example, from their own scientific experiments, or from topical sources such as newspapers or magazines. In contrast, the use of artificial figures, including most old statistics which are effectively artificial, is far less effective.

I hope therefore that teachers will supplement, or even replace, the data in these problems with data that are more topical and relevant to their pupils.

The data in these problems are neither as complicated nor as simple as they might be. In general, the data should be as complicated as the solvers can cope with, giving them something to get their teeth into, and maximising their sense of achievement.

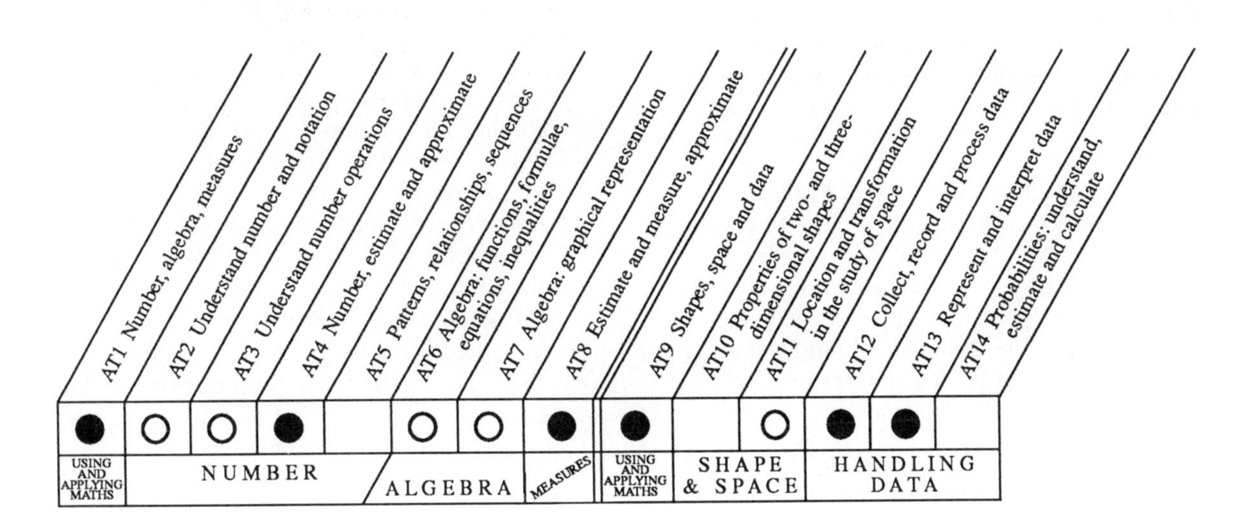

23 Making sense of data

Comments and suggestions

A1 General problems such as this emphasise the impossibility of saying much about a set of data by summarising it in one number only. Compare problems **D3–5** which emphasise the same idea by asking how much you can know by working backwards from just one figure.

Does the purpose for which a representative height is demanded make a difference? How could the data in the list be summarised in one English sentence? How could it be summarised in two numbers, rather than one?

Are some of the data so exceptional that they can be ignored, or are the extreme cases especially important?

Note that the problem does not say whether the height to be chosen should be one of those listed.

A2 The initial 4 looks like an odd one out. Whether it should be included or excluded depends on the context, which is not stated. If these are prices of items which have to be purchased, the 4 is very relevant and lowers the average price markedly. If they are the results of a scientific test, the 4 might be disallowed as an obviously freak result. A sound response to the problem could be, 'It depends!'

A3 A powerful intuitive idea of an average is that it is what you get by literally taking a bit off here and putting over it over there,... until the piles, in this case, are as nearly equal as possible. This interpretation also relates directly to smoothing a set of data.

When is it possible to make all the piles equal in height? Is it always possible ensure that the difference between any two piles is never greater than one unit?

How can the arithmetical average, when it is not an integer, be read off from the final arrangement of nearly equal piles?

A4 The wording points directly to the significance of the total cost, which is not necessarily obvious when the data are smoothed by taking a bit here and adding it there.

Can the average of all four prices be calculated by finding the average of them in pairs, and then finding the average of the two averages?

——— ◆ ———

B1 This illustrates the kind of data which is within reach of most pupils. Far from being too complex, it gives them more to think about.

Most pupils have a strong intuitive idea of a trend. The real problem here is putting the idea into words, or deciding how to justify an answer to the question in arithmetical terms.

How should the phrase 'on average' be interpreted here? Does it mean, literally, that an average should be calculated, or is its meaning more general?

B2 The human eye can hardly fail to see the dots as rising to the right. Are they, however, really rising in a rough straight line, or in a curve? In other words, are they on average rising steadily, or is the rate of increase itself rising or falling?

If the edge of a ruler is used as a variable straight line, how should the ruler divide the set of points? Should there just be as many points above as below the line, or should the distance of points from the line be taken into account.

In what sense is a straight line showing the general trend, an 'average' line? This set of points has an average point, a centre of gravity: is there any reason why the trend line should pass through, or even near, the centre of gravity?

——— ◆ ———

C The problem assumes the idea of smoothing, and suggests an important idea. Although smoothing shows up trends, it also loses some of the original information.

In what sense is a scientist, for example, studying the data from an experiment, or a market researcher looking for patterns in the results of survey, happy to lose information in this way? When is the trend the real and useful information, and the individual data a mixture of the genuine trend and random factors?

——— ◆ ———

D1 This and the following problems force pupils to think hard about what average, median and mode mean, and also suggest strongly that these figures give very little information about the original data. An infinite variety of different sets of data can have the same average, or even the same average, same median and same mode, as well as the same range,... .

Why then are these measures of any practical importance? Firstly, because the particular practical conditions may make one of them appropriate. For example, a shoe shop manager will be very interested indeed in a small number of popular sizes. Secondly, because in so many practical situations the distribution is rather well-behaved and symmetrical, so the person who appears to know, say, only the average, in fact knows a lot more than this.

Does the problem imply anything at all about the number of data in the solution?

Given a set of data within a given median and mode, for example,

1 2 3 4 5 5 6 7 8 9

is it possible to adjust the average to any target figure by inserting one extra number?

Given a set of data with a given average, can the median and mode be changed at will, by adjusting the data so that the average does not change, by adding a bit here and taking off the same bit there...?

D2 How can the numbers in a set be adjusted to make the median twice the mode? How can they then be adjusted to make the average twice the median?

How could a set of numbers be adjusted so that its average is unchanged, but its median becomes one half of the average?

Is there in general any connection at all betwen the figures for the average, median and mode? It is possible to say anything at all about one of these figures, if you know the other two?

D3–4–5 In each case an appropriate answer might be, 'Very little indeed!' Without extra information, such as some practical context, the numbers in the set can be chosen with a high degree of arbitrariness.

What other measures of a set of data might provide more information? What short descriptions in words, such as 'symmetrical' might provide more useful information?

———◆———

E What are the speeds on the different routes? How accurately can the speed of an aircraft be estimated from this kind of data? Is there any evidence that different types of aircraft are used on long and short routes?

How many different factors might be involved in the variation in flight time, which cannot be separated from these data alone?

24 Averages

'On average' is a very common everyday expression, and everyone has a strong intuitive idea of what it means. This makes it easy and motivating to pose problems to pupils about averages. Such problems seem 'natural', and yet they force pupils to think very carefully about what the mathematical definition of an average means and what its consequences are. They will discover that, as so often, the mathematical concept and the everyday concept are very close in some respects and surprisingly different in others.

Apart from positive everyday associations, averages are extremely important in mathematics. The relationship between the arithmetical average and ideas of proportion, balance and 'centre' (another common everyday term with many mathematical interpretations) is very striking. An average is a linear function and therefore shares the properties of linear functions. An average is also effectively a centre of gravity, and conversely, so that averages are easily interpreted in simple mechanical devices.

An average is also a sum, and therefore at a higher level closely related to integration. Finding the average value of a function over an interval, for example of speed against time, is one approach to integration.

In contrast, the mode and the median have far fewer properties and are of much less mathematical interest. One way to contrast the difference is to write computer programs which will calculate the average, median and mode of a set of input numbers!

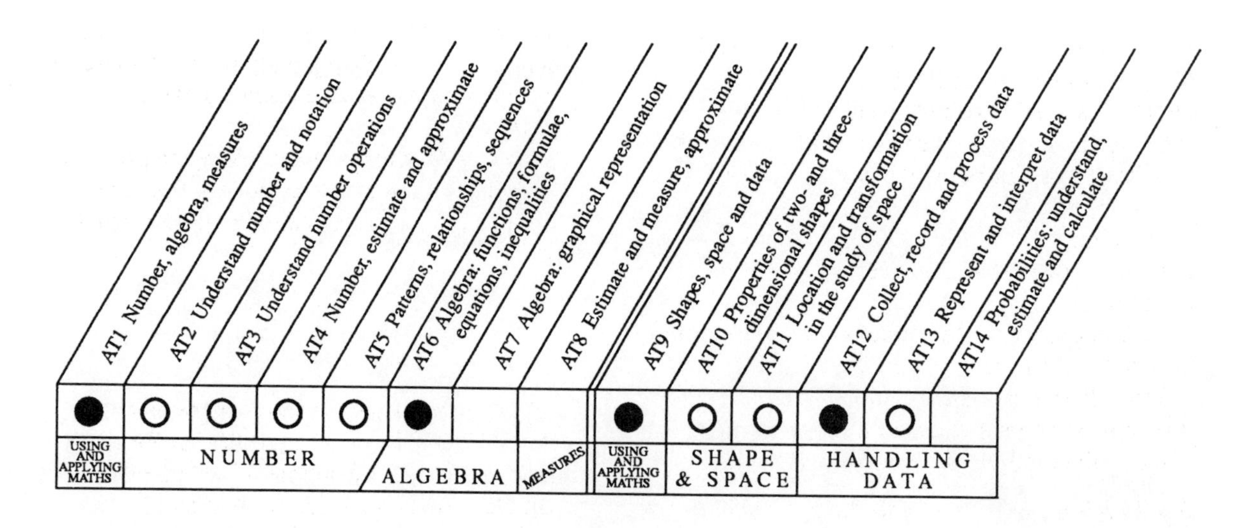

24 Averages

Comments and suggestions

A1 These basic properties of the average help to explain why it is so important. What happens to the median and the mode under the same conditions?

What other operations, performed on each number in a set, will have the same effect on their average?

A2 Pure calculation gives an unilluminating answer. The average depends on the total of all the numbers in the set, and the number of numbers in the set. Does the total, in this case, change significantly? Does the number of numbers?

What would be the effect if the additional value was not 21, but, say, 21 000?

How can the average of a set of numbers be roughly estimated if several of them are far larger than the others?

A3 How can the average of a set be changed by a large amount, by the inclusion of one extra number? It is intuitively plausible that an extreme value will do the trick.

What extra number will have least effect on the average?

What happens to the average if the new numbers included are, on average, greater than the average?

What number could be included which would double the average?

How do the median and mode of a distribution change when additional numbers, equal to the median or the mode, are included?

A4 This is another way of looking at the idea that the average is what you get by smoothing the numbers out, by taking bits off the larger numbers and adding them to the smaller numbers. (There is no matching process for the median or mode.)

If a sequence of numbers is smoothed by replacing each number by the average of the number and its successor, with the last number linked to the first, what will happen to the average?

A5 Under either, or both, of two different conditions. The solution of problem **A3** is a hint here; when new numbers equal to the original average are included in a set, the average does not change.

This problem and **A3** are difficult to think about without any use of algebra. They are on the borderline, as it were, between problems which are best solved by careful thinking and insight, with no formal manipulation at all, and problems which are easily solved algebraically, although formal manipulation may lead to, and certainly generally requires, little or no insight at all.

Algebra is so powerful partly because it obviates the need for so much insight, but this is also its weakness, especially for pupils.

A6 Compare problem **A4**. This method of finding an average can be thought of as reducing every entry by 40, finding the average of the new set of numbers, and then restoring the 40.

This problem does not require any formal understanding of negative numbers. Pupils unfamiliar with them can discuss what it could or should mean to add '17 more and 23 less'. In this way, a problem like this becomes a means of introducing and motivating concepts of negative numbers, rather than an exercise in their application.

Why was 40 chosen as the reference point? Is this the easiest number to choose, if the object is to calculate the average mentally, rather than by calculator?

———— ◆ ————

B1 What special case shows that the answer is certainly 'Yes!'? Does it help if the cardboard is initially cut into, for example, 12 equal pieces, so that the solver has effectively to divide 12 into three numbers, one of which is the average of the other two?

This problem depends on the special fact that the average of *two* numbers is, as it were, 'equidistant' from each of them. This is not true of more then two numbers.

B2 If **B1** is not solved first, then pupils will probably need to experiment with actual pieces of cardboard, and solve several specific problems, to maximise the chance that they will see the general solution through insight.

How many solutions are there to this problem?

———— ◆ ————

C1 This problem and the next show how problems about averages can be asked about geometrical objects or situations.

Compare questions such as: 'What shape is the average triangle?' Such questions are plausible because the idea of an average is so intuitively strong. They often have answers only after the original question has been adapted or refined to take into account its ambiguity and vagueness. How is the 'shape' of a triangle to be measured for the purposes of finding an average?

What is the average number of edges in the commonest tessellations, such as squares, hexagons or triangles?

What is the largest possible average number of edges in a tessellation?

Given a number, say, $3\frac{1}{2}$, is it possible to design a tessellation with an average of $3\frac{1}{2}$ edges per tile?

Can two tessellations with different patterns, made from different tiles, have the same average number of edges?

How does the average number of edges depend on the boundary of the tessellation? Is it reasonable to assume that the tessellation goes on for ever and has no boundary? In which case, how do you find the total number of edges, in order to calculate the average?

C2 What are the average numbers of edges of the faces of the regular and semi-regular polyhedra?

The regular dodecahedron has an average of five edges per face. A modern football which is in the shape of a truncated icosahedron with a mixture of pentagonal and hexagonal faces, must have an average of between five and six.

What is the largest proportion of the faces of a polyhedron which can have six edges?

Take a plastic ball and start to draw a tessellation on it. How do the possible tessellations on a sphere compare with plane tessellations?

◆

D1 Compare Unit 18 problem **A5** (p. 126). The present context suggests that the answer is, in some sense, their 'average'. In general, all problems about balancing and centres of gravity can be thought of as problems about averages.

Experiment with weights and a marked rod will confirm the solutions. How can an experiment best produce the effect of a rod which weighs 'almost nothing'?

D2 How should allowance be made for several numbers in the set being equal?

How can the instrument be designed to average as accurately as possible both sets of small numbers, say between 1 and 10, and sets of much larger numbers, say between 10 and 1000?

How small can the experimental error be?

◆

E1 Compare Unit 26 (pp. 180–2). What is the average throw of a dice in a small number of trials? How close is the experimental figure to the expected figure?

E2 Is it fair to say that the average for two dice thrown together must be double the average for one dice? Would the average for 3 dice be three times the average for one dice?

Where would the average plausibly be located if a scatter diagram were drawn for the different ways of throwing two dice together?

F1 Contrary to expectation, for many pupils the idea of the 'centre' or average for a 2-dimensional set of points on a plane seems easier to handle than for a 1-dimensional set of points on a straight line. This is perhaps because the idea of the centre of a surface, or of a solid shape, is common in everyday life.

How might an average be judged by sight? How might an average be calculated? Is it sufficient for a calculation to consider the weights and heights separately? How close is a calculated solution to the average picked out by visual inspection and judgement?

It is easy to make a scatter diagram for the height and weight of pupils in a class, or for any other pair of characteristics. How close is a calculated average to the pupils selected by a vote of all the class as most average in height and weight? Does the average mean anything if the scatter diagram shows the boys and girls in two overlapping clusters?

What would a scatter diagram in 3 dimensions look like? How could a scatter diagram in 4 dimensions be analysed?

25 Combinations and permutations

Problems in combinations and permutations can be traced back at least as far as the correspondence between Pascal and Fermat on problems of probability and Pascal's book on the arithmetic triangle which bears his name, though he was not its inventor.

Such problems are often easy to state, requiring no mathematical language or concepts and no prior knowledge of mathematics, like the best puzzles, yet their solutions are not as simple as they seem. On the one hand, they share the common heuristic feature that without a carefully thought-out plan, without a system or method, the solutions will be best be unproven and at worst incomplete and full of repetitions. On the other hand, the answers are often surprisingly large. If the solutions are expressed as functions of the original data, then these functions, for example $n!$, increase very quickly indeed.

Pupils who, quite naturally, have initially no conception of either of these features do not realise what they are letting themselves in for, and can easily get bogged down in unorganised lists of possibilities, in which there is no pattern, no regularity, and very little mathematics.

Pupils need one of several classic problem-solving heuristics. One strategy is to start with very simple cases, replacing the data in the problem with smaller data, solving the problem for a sequence of simpler case and so working up to the more complicated case. Very often, the pattern in the sequence of simpler problems can be spotted, and the answer to the original problem induced. Being more or less confident of the answer, pupils may find a solution easier to work out.

A second approach is to break the problem down into several smaller problems, by considering different cases. A third is to tackle the original counting problem directly, but listing the possibilities according to a plan which guarantees that they are all included, just once each.

All these heuristics are illustrated by the problems which follow.

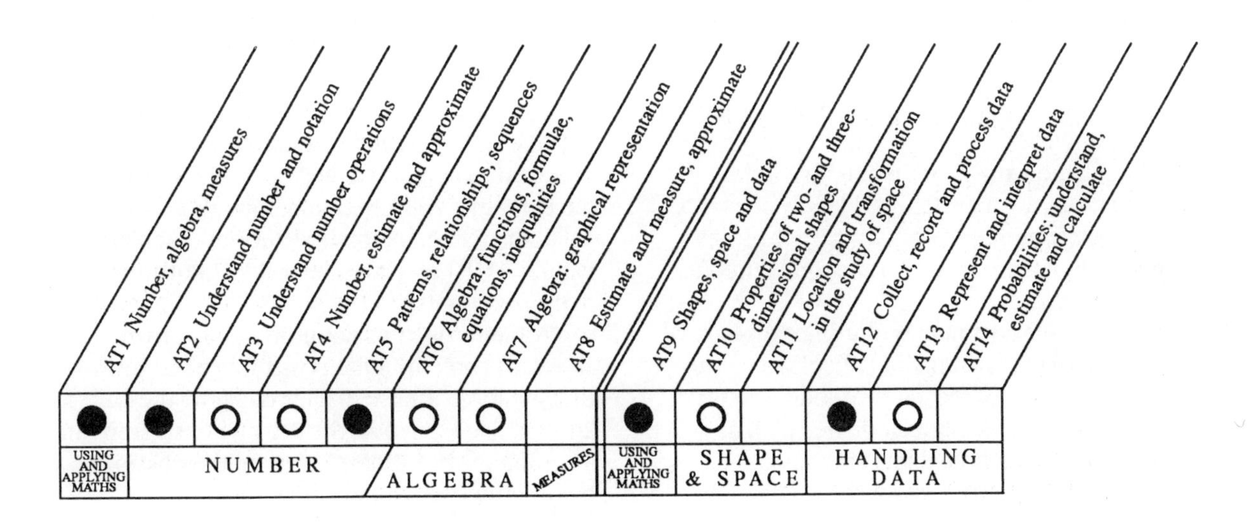

25 Combinations and permutations

Comments and suggestions

A1 How many combinations would there be if there were only three-cylinders, each marked from 1 to 4?

The problem would be essentially the same if each cylinder showed the letters A–J. The use of digits makes it easier because each sequence is then a number between 0000 and 9999. Is there a one-to-one correspondence between positions of the cylinders and these numbers? What movements of the cylinders will generate the entire sequence in order?

What is the connection between a combination lock with cylinders marked from 0 to 5, or from A to F, and powers of 6 and numbers written in base 6?

What is the connection between the combination lock of this problem and the cylinders of an old-fashioned desk-calculating machine, or Pascal's original calculator?

A2 How many combinations would there be if a postcode consisted of letter-digit-letter, only? How could these be listed systematically?

In what respect is the set of postal codes like the positions of combination lock in which there are 26 positions of the first two cylinders, 9 for the third, 10 for the fourth, and 26 for each of the last two?

Roughly how many of the possible letter combinations are actually used? How does the total possible number compare with the total population of the country, or the total number of households?

What are the advantages of using a mixture of letters and digits, as in the old telephone codes? In the United States, postal 'zip' codes are two letters indicating the State, followed by five digits. In Canada, three digits and three letters alternate. Could these systems be used in this country?

A3 An important idea in all these problems is that if the possibilities are genuinely independent, then they are combined by multiplication, and their order makes no difference. In this case, the choices of trousers, jackets and suits are not independent. There is a possibility also that on a hot summer's night he might not wear a jacket (or suit).

Can the problem be solved by considering the total number of combinations of shirt-and-tie, trouser-and-jackets-or-suits, and shoes, and so reducing it to the problem of combining three sets of possibilities only?

Can it be solved by considering the combinations of suits-and-shoes, or jacket-trousers-shoes, and shirt-and-tie and then combining two possibilities only?

A4 This problem is much like the previous problem. In particular, just as Pete Perfect's large choice of ties boosted the number of combinations he could wear, the choice in this restaurant is boosted by the relatively large choice of dressings.

How could a picture of all the possibilities be designed?

B1 The **B** problems all invoke the order of the items, and therefore involve the factorial function $n!$ One way to introduce the rapid increase of this function is to ask pupils to calculate 1!, 2!, 3! and 4! and to predict how large, for example, 10! will be. For pupils who have not met the function before, the result is very surprising. (As an indicator of much higher values, 450! has 1001 digits, and 1 000 000! has 5 565 709 digits.)

Problems like this naturally break down into simpler problems if the possibilities for, say, the first position are listed. This could be the start of an organised list of all the possibilities. How can a complete strategy for listing all the possibilities be described accurately in words?

B2 What is the connection between this problem and **B1**? In contrast to **A1**, the presence of digits rather than letters is not a help, apart from observing that almost all the 1 000 000 000 combinations from 000 000 000 to 999 999 999 are excluded because they contain repeated digits.

(What is the probability, in the light of this problem, that a nine-digit number contains a repeated digit? What is the probability than an n-digit number contains repetitions?)

This is an ideal problem for solution by constructing a sequence of simpler problems, starting with two-digit numbers using two digits, three-digit numbers using three digits,..., leading to the sequence of factorials. The initial 1 might be added after examining the sequence from 2 onwards. The idea that you can even reasonably ask for the number of one-digit numbers using the digit 1, one at a time, may seem weird to pupils! To mathematicians it is an important general concept that you almost always can ask such curious questions and get answers that fit the pattern.

Alternatively, the sequence of problems can be constructed with one-digit numbers, using the digits 1 to 9, two-digit numbers with the same condition, three-digit numbers, and so on. This leads to the different sequence:

$9 \quad 9\times 8 \quad 9\times 8\times 7 \quad \ldots \quad \ldots \quad \ldots$

B3 Typically, with so many letters (26) it is impossible to list all the possibilities (at least without the aid of a computer program). Fortunately this problem is effectively solved by any of the plans of strategies used to solve **B1** and **B2**.

In contrast to 9!, which can be calculated on an ordinary eight-digit calculator, 26! is far too large. How might it be calculated without the use of a computer? How might it be calculated using a computer, considering that the computer language used will probably not copy with such large numbers without extra programming? How might its value be estimated?

How, in principle, could the number of arrangements in order of, say, 100 objects be calculated?

B4 According to Plutarch, Xenocrates (396–314 BCE) calculated the number of syllables that could be made from the Greek alphabet as 1 002 000 000 000. If this story is true it is the first solution of a difficult problem in combinations and permutations.

The number of legitimate three-letter words in English is smaller than word-game players would like. Many plausible-sounding words are not in the dictionary.

There are several possible problems here. Should Y be counted as a possible vowel? Can a three-letter word have two vowels? (Presumably not three?!) Is no attention to be paid to possible pronunciation?

How can the problem be tackled in stages?

C1 Another example of a problem which is well tackled by starting with a smaller number of elements. What is the sequence of handshakes if there are 2, 3, 4, ... directors?

How many extra handshakes are made if another director arrives late at a board meeting, after everyone else has shaken hands?

Alternatively, each director's handshakes could be listed, systematically. How can the solution be calculated by a formula based on the number of directors?

How does this problem relate to the problem of selecting two objects from a set of 10, or n?

C2 What is the relationship between this problem, and the previous problem? Why are the solutions not the same?

How many extra diagonals are added when an extra vertex is added to the polygon?

C3 Notice that the rectangle is divided into smaller rectangles, so the question of whether squares count as rectangles does not arise.

At first sight, this problem is complex and requires the addition of the totals for a variety of different-shaped rectangles. Is this the only way to solve the problem?

How many rectangles are there if rectangles are divided into fewer equal parts along the sides?

How many rectangles are there in each of each of this sequence of rectangles?

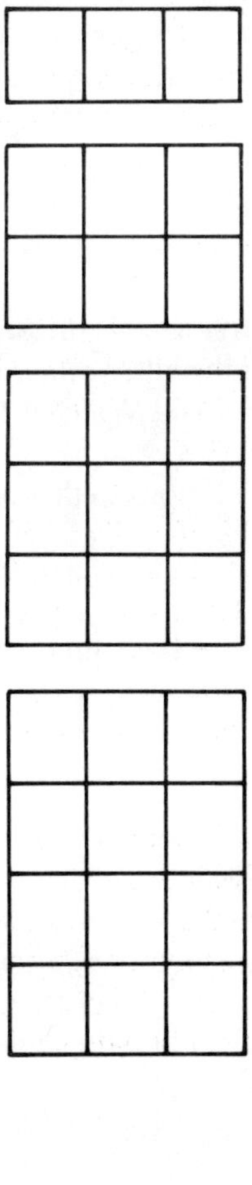

C4 It would seem natural to get some idea of what the solution might be by counting the triangles in these figures.

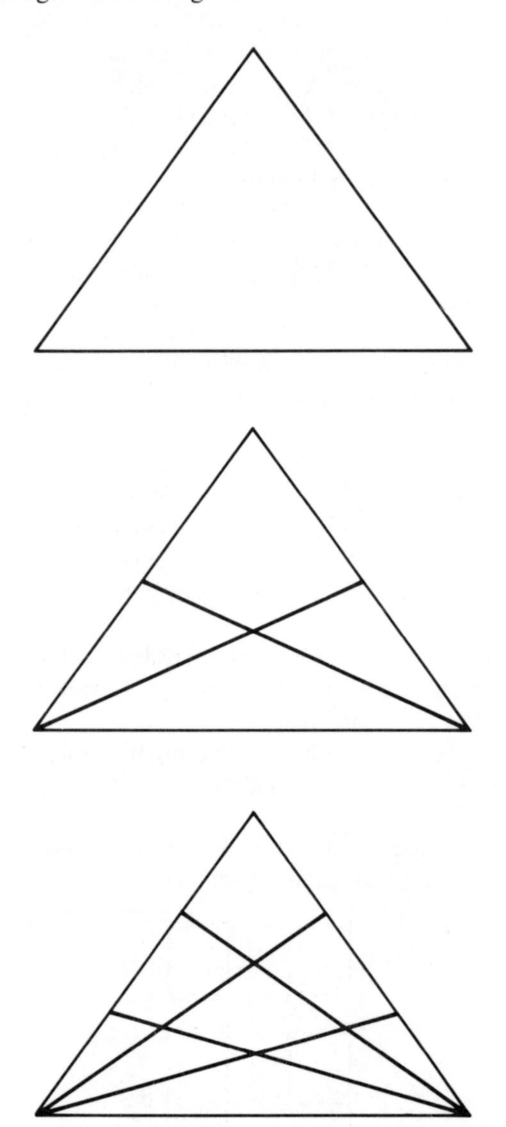

To count the triangles in the original figure, it may help to spot that all the lines in the figure do not have the same role. Which is the odd-line-out?

Why are the solutions to these problems such round numbers?

26 Find the false dice

The title of this set of problems includes the word 'false' for two reasons. It is emotionally charged and therefore adds to the affective significance of the problem. It is naturally more interesting for most people to find mistakes and errors than it is to discover or confirm that everything is perfect. Thus if several groups or individuals are tackling this topic, a few of them might have supposedly sound dice, and the others have dice which are definitely defective, without either group knowing the status initially of its own dice.

Secondly, asking to what extent and in what way a dice is biased and asking for the probability that a recorded sequence of throws came from a biased dice are actually deeper questions than simply testing for correctness.

Polyhedral dice are available from shops selling war and fantasy gaming equipment, but they are expensive, and dodecahedral and icosahedral dice roll a long way before stopping and can be difficult to read.

Cubical dice are only one kind of shape that is traditionally used to determine odds. Any flat object can be used as a two-sided coin or dice, many kinds of spinners and tops can be made, and other polyhedra such as the six-sided prism used in a traditional children's cricket game can be used.

If children make their own dice, then an extra element enters; who can make the most accurate dice, and how can you tell?

A computer is ideal for simulating experiments, though of course pupils will have to decide in advance what it means for a dice to be fair, in suitable mathematical and programming terms, and how a simulated dice might have bias programmed into it. These are themselves very interesting problems.

In a different direction, a computer program can be used for listing combinations and permutations, which also poses difficult and interesting problems in mathematics and programming.

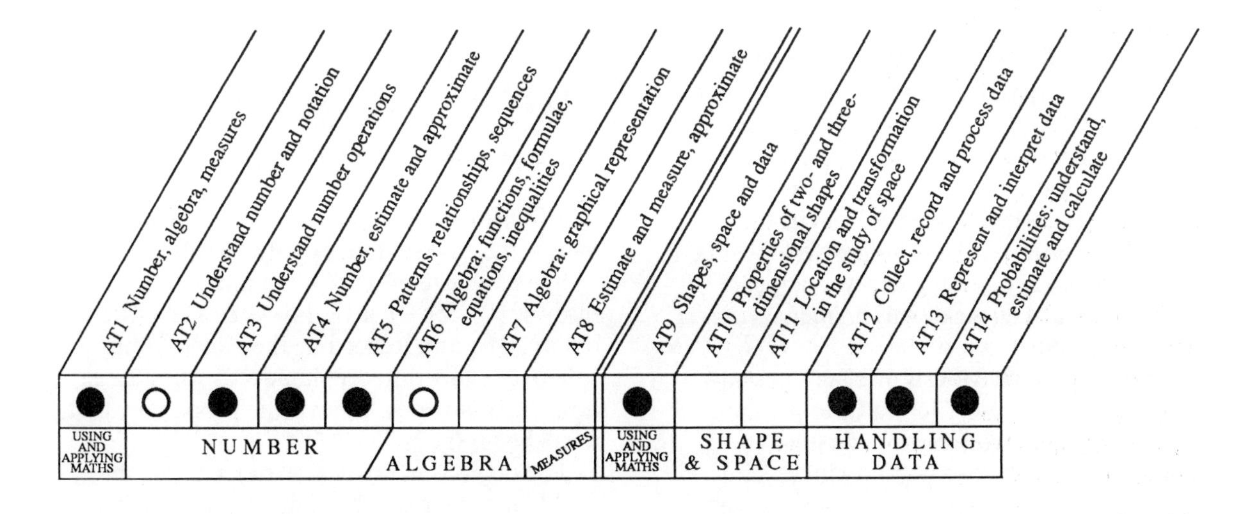

26 Find the false dice

Comments and suggestions

A The **A** problems here are richer and more general than the **B** problems. Indeed, some of the solutions to the **B** problems are needed in order to tackle the **A** problems. If pupils start experimentally by throwing their test object a large number of times, equipped only with the simplest intuitions, for example that heads and tails ought to fall 'on average' the same number of times, then failure of their results to entirely match their expectations will force consideration of some of the **B** problems.

The **A** problems may therefore be thought of as extremely natural (and practical!) problems which motivate the **B** problems, several of which are much more abstract and less obviously motivated. In turn, the **B** problems help to solve the original **A** problems.

A1 On the one hand, it is natural to suppose that heads and tails, or the six faces of a die, will fall equally often. On the other hand, it would be astonishing if they fell in sequence, for example H—T—H—T—H—T... or 1—2—3—4—5—6—1—2—3—4....

What extra conditions will ensure that sequences such as these do not count as random?

The actual sequence of events will in some sense be random when looked at closely, but more ordered when looked at from further away. How ordered? How long a view must be taken for the expected regularity to appear? Problems **B1–4** all tackle this point in different ways.

If the object is obviously 'not fair', perhaps because it is unsymmetrical (a short prism or an almost cubical rectangular box might be practical examples, or a pebble with several flattened sides), is it possible to estimate the distribution of the different ways it can fall, and hence predict which sequences are likely or unlikely?

A2 How many times do you need to throw a pair of objects in order to be confident of your result? If one object is consistently fairer than the other over the first 50 throws, can you be certain that it will also be fairer over the next 50?

Study of **A1** would suggest that the answer to the question is 'No!'.

How large would the difference in the distribution of throws be before you could reasonably conclude that they were not equally fair?

How could 'fairness' be measured, so that you could then say, 'The fairness of this dice is such-and-such...'?

A3 What total scores would you expect to see most frequently when two dice are thrown and totalled together? This is problem **B6**, below.

If you did not get the pattern of total you expected, would it be possible to say which of the dice was unfair, or only that one of them was unfair? Or both of them?

If the expected distribution of totals does appear, can you be confident that both dice are fair, or is it possible that they are both unfair but in 'opposite directions'?

Is the strategy of this problem, adding the totals to save time, really a time-saving way of comparing the fairness of two dice?

◆

B1 Experimenting with several hundred throws will suggest the range of splits between Heads and Tails that is likely to turn up in practice. Why is the split seldom exactly 50–50?

In a sequence of (say) 300 consecutive throws, 100 consecutive throws can be picked out in 201 ways. Recording the Heads–Tails split over many different hundreds will suggest how the split varies. What is the probability that it will be 50–50?

The obvious heuristic for investigating this problem further is to consider sequences of a very small number of throws, looking for patterns which might possibly be continued to much longer sequences of throws.

How many Heads and Tails can be expected in a sequence of only four throws? How likely is it that Heads and Tails will be split 50–50?

B2 What does 'how many times' mean? What could it mean?

Is there any reason why Heads should be followed by Heads rather than Tails? Or why Heads should follow Heads rather than Tails?

If instead of looking at a long sequence and counting the occurrences of Heads and Tails, you look at all the possible short sequences, how many are there? In how many of them do two Heads appear following?

If three Heads appear in sequence, does this count as an example of 'two Heads in sequence'? Or perhaps as two examples of a 'two Heads in sequence'?

This problem is closely related to **B3–5**.

B3 This problem refers specifically to a limited sequence of four throws. Is this an entirely different kind of problem, therefore, from **B1** and **B2**? Would it be safe to simply examine a long sequence of throws, and write down all the different sequences of four consecutive throws that occurred?

The problem refers to a single coin being thrown four times: what is the difference between this problem and the problem of four coins being thrown once each, in sequence? Or simultaneously?

What systems (an appropriate word when talking of odds!) can be used to make sure that a listing of all the possibilities contains each possibility exactly once?

B4 This problem could be tackled experimentally, or by considering the possibilities logically, or both approaches could be used and the results compared.

Because the problem asks specifically about two Tails out of eight throws, one heuristic would be to study the probabilities of getting two Tails out of two throws, three throws, four throws,...looking for a pattern.

Will the answer be the same as the probability of getting one Tail and three Heads out of four throws?

Is it necessary to list all the possible outcomes of eight throws? What is the significance of this problem: 'Eight pupils are arranged in a line. In how many ways could you select two of them?'

B5 This problem is closely related to the previous three problems. Indeed, all these problems are, as it were, different ways of asking essentially the same questions, although this problem does not refer in any way to chance or probability but is entirely, as it were, static, a problem in geometrical arrangements in a straight line only.

Typically, there is no mathematical difference between such static problems and dynamic problems about throwing coins or dice.

B6 This problem could also be solved by a practical experiment, or by logic. A comparison of the two approaches may suggest answers to some of the questions raised by problem **A3**, above.

How can all possible totals be displayed as elegantly as possible?

What is the difference between throwing one dice twice and adding the totals, and throwing two dice simultaneously?

What distribution of possible totals will occur with unusual dice – for example, the tetrahedral or icosahedral dice with 4 and 20 faces respectively used in some games?

How can a dice be designed so that certain totals are most likely? Can a cubical dice be marked so that all possible totals are equally likely?

B7 The results of throwing three coins can be represented in a three-dimensional diagram, as in **B2** above. Can the same be done for the results of throwing three dice?

How else can the possible totals for three dice be displayed clearly? Is it necessary that they be displayed in order to solve this problem?

What happens when different kinds of dice are thrown together, for example, a cubical and a tetrahedral dice? Or one cubical, one tetrahedral and one icosahedral? (This problem presumably assumes that the dice are normal cubical dice.)

27 Combinatorics

Most of these problems ask for a task to be achieved, offering a very basic motivation. It is often possible to reach the goal, or persuade oneself that the goal can be reached, by trial and error. Such partial solutions invite a more careful analysis which describes exactly what must be done and why it works.

Subsequently, they all offer further challenges which are usually more mathematically demanding. In how many ways can the goal be achieved? In how few moves? What happens if the initial map, network, or position is changed? When can the goal not be achieved? And so on.

The mixture of traditional problems and the ideas of several particular and famous mathematicians who first thought of them long before combinatorics existed as a branch of mathematics in its own right, is very striking.

Typically, they might have occurred to anyone, and had originally the nature of recreational puzzles, amusing and tantalising no doubt, but with no deeper significance. It is only since the development first of topology and then of graph theory and finite mathematics generally that they can now be seen in perspective as hints of explosive developments to come.

The combination of simplicity in the posing of several of these problems, and the unsuspected depths to which they lead, means that a wide variety of pupils can tackle these problems at their own level, without exhausting the possibilities in any of them.

At the same time they are gaining experience in a type of mathematics which is becoming more and more important in applications, including physics and chemistry which have been the traditional domains of calculus, and which although given little prominence on syllabuses, provides very rich opportunities for the problem-solving (investigative) work that is rightly demanded by the GCSE and the National Curriculum.

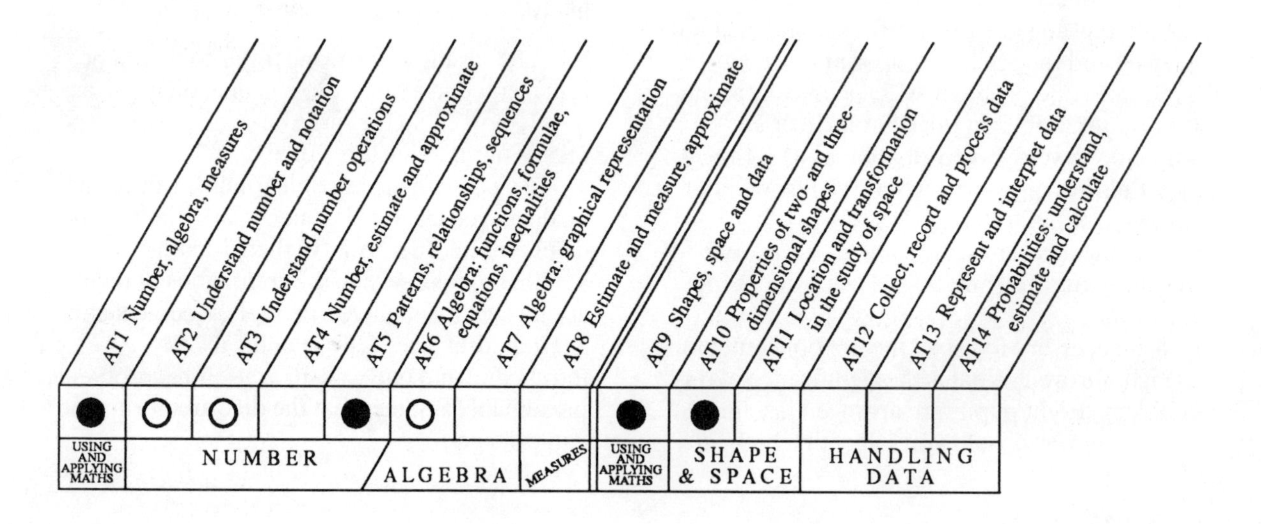

27 Combinatorics

Comments and suggestions

A An important general concept in problems like this is the idea that the solution is (almost) sure to be symmetrical. Even if there are strikingly non-symmetrical solutions, it would be foolish not to look for symmetry first, in order to get some general idea of the possibilities.

One start might be to see what combinations of points-on-lines and lines-through-points there are in some common figures, in order to get some feel for the problem.

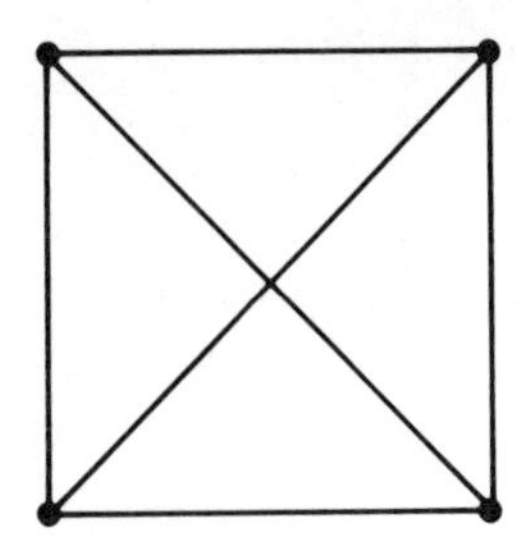

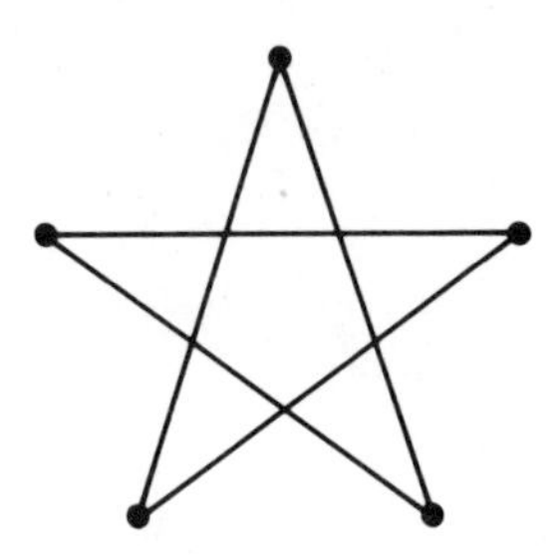

This square has 6 lines, 4 points and 3 lines through each point and 2 points on each line, if the centre is excluded. Is it allowable to exclude some points of intersection? What other lines might be added to give different combinations of points and lines?

The pentagram star has 4 points per line and 2 lines per point. What possibilities do other polygons offer?

A pentagon does not naturally have 3-fold, let alone 9-fold symmetry. What about a star 9-gon? What about a triangle which already has 3-fold symmetry?

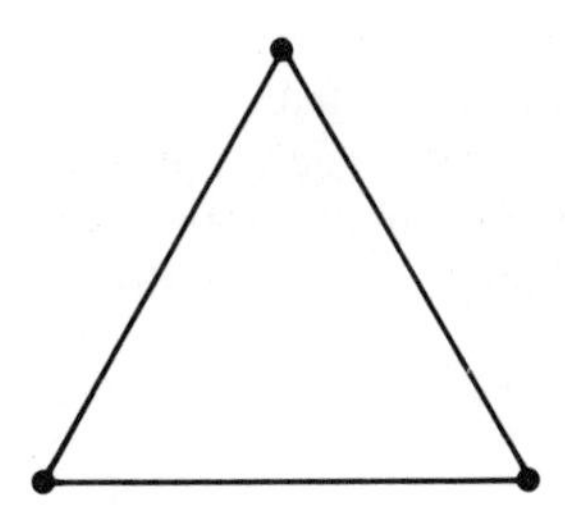

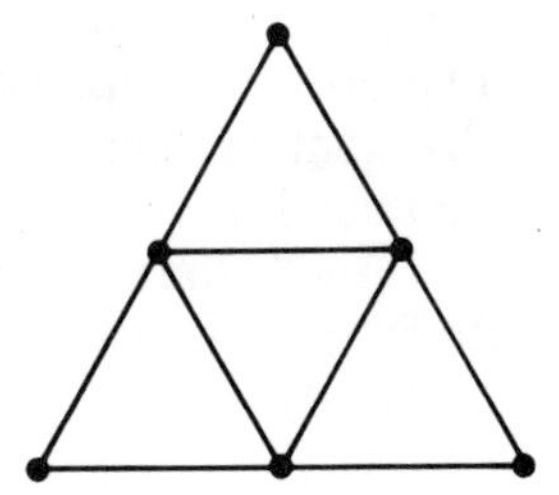

The first figure, an equilateral triangle, shows 3 points, 3 lines, and 2 points on every line, and 2 lines through every point. The second, with an extra equilateral triangle added, has 6 points and lines, and 3 points on three of the lines, but only 2 on the other 3 lines. Can further points be added to get rid of this imbalance, and even to match the specifications of the problem?

◆

B This is one of many games and puzzles which start with a network. Many of them (perhaps surprisingly in view of the simplicity of the idea of a network and the ease with which problems about networks can be posed) are of great mathematical importance.

Do the rules say that the attacker must mark his first edge starting from S? Or can any edge be marked at a turn? What difference would this decision make to the attacker's chances of winning?

Games are typically complex situations in which playing the game is essential to develop some feel for the possibilities. Who wins most frequently in practice in this game? If the last few moves of a game are retraced, could either player have improved on their subsequent moves?

What simpler boards are easier to analyse? Are there patterns of boards in which one player can obviously win by force?

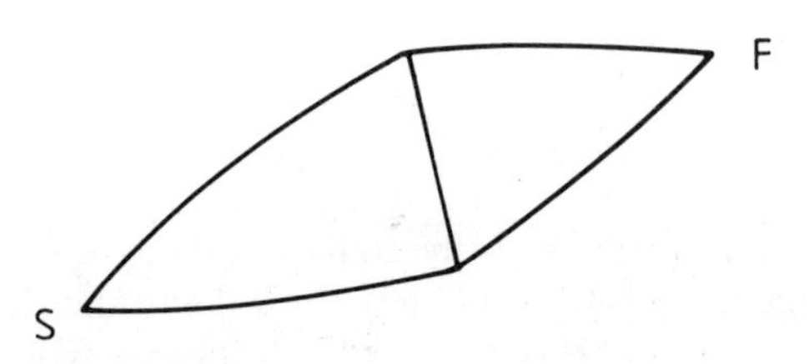

Which line should the attacker mark in order to be sure of eventually joining S to F in this simple network?

◆

C Lucas originally wrapped this problem in a fantastic story about a Hindu temple within which the priests continually transferred a pile of 64 golden discs from one peg to the other; when the task was completed, the world would come to an end. The Brahmin priests would take more than 600 000 000 000 years at one move per second, all day and every day.

What patterns do the discs make as they move from peg to peg? To which peg should the topmost disc be initially moved? Where does the idea of parity come into the mechanism of this problem?

How can 2, 3, 4 or 5 discs be most efficiently moved?

What is the best notation for indicating the moves of the discs?

How can the given problem be broken down into two simpler problems?

How much faster would the task be if there were 4 pegs available, rather than 3?

——— ◆ ———

D The humble 3 × 3 magic square illustrates several important general concepts. The diagram of 8 cells is highly symmetrical, and so is the sequence of digits 1 to 9. What kind of symmetry can be expected in a solution?

What will the sum of each row and column be?

Is it possible to predict, or calculate, the value of the central cell?

Placing numbers in the cells at random is not an easy way to a solution.

What placements can be eliminated as definitely impossible?

How does the symmetry of the 4 × 4 square aid a solution? In what way might a solution resemble a solution to the 3 × 3 square?

——— ◆ ———

E This problem is due to Ernest Dudeney. What difference would it make if the farmer had a free move at the start? What difference would it make if the farmer and the pig had different moves, moving for example like knights in chess?

Trial and error will suggest that the farmer can never catch the pig. How could this be proved, in the sense of being argued so simply and clearly that the conclusion was obvious, even without trial and error?

——— ◆ ———

F This is a version of the *pigeonhole principle,* which seems very simple indeed, but nevertheless has many elegant and important applications. It is related to the idea that if several piles are of unequal height, then at least one of the piles is taller than the average height.

To prove that a situation must be so, it often helps to explain why any other situation is impossible.

——— ◆ ———

G Like **F**, this can be an example of the pigeonhole principle. To find a counter-example the points need to be spread as far apart as possible. How widely can five points in a square be spread?

The same idea can be applied to an equilateral triangle. In this figure, why must at least one pair of five points be less than 1 unit apart?

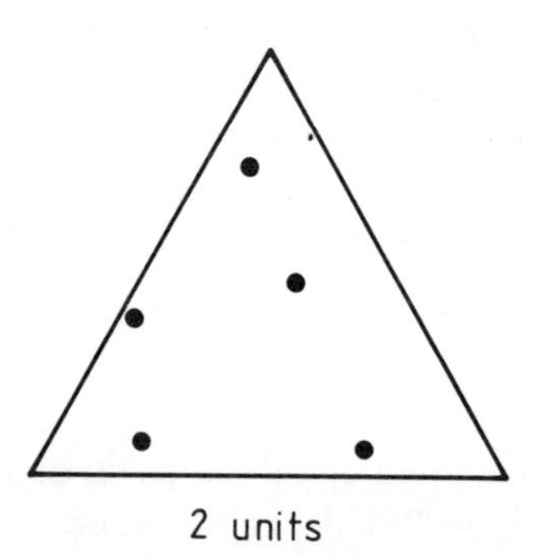

How widely can, for example, eight communication satellites be spread above the surface of the earth?

——— ◆ ———

H A venerable puzzle type, but also another example of the pigeonhole principles. If there are more inhabitants of a town than any inhabitant has hairs on his or her head, is it true that there is a least one pair of inhabitants who have exactly the same number of hairs on their heads?

——— ◆ ———

I This is a good example of a very practical problem, which is much simpler than similar problems about, for example, the flow of